MAKING ROUTES

MOBILITY AND THE POLITICS OF MIGRATION IN THE GLOBAL SOUTH

Edited by
Gerda Heck
Eda Sevinin
Elena Habersky
Carlos Sandoval-García

Foreword by Ibrahim Awad

The American University in Cairo Press
Cairo New York

First published in 2024 by
The American University in Cairo Press
113 Sharia Kasr el Aini, Cairo, Egypt
420 Lexington Avenue, Suite 1644, New York, NY 10170
www.aucpress.com

ISBN 978 1 649 03317 8

Library of Congress Cataloging-in-Publication Data

Names: Habersky, Elena, editor. | Heck, Gerda, 1968- editor. | Sandoval García, Carlos, editor. | Sevinin, Eda, editor.
Title: Making routes : mobility and the politics of migration in the global South / edited by Elena Habersky, Gerda Heck, Carlos Sandoval-García, Eda Sevinin.
Identifiers: LCCN 2023016656 | ISBN 9781649033178 (hardback) | ISBN 9781649033185 (epub) | ISBN 9781649033192 (adobe pdf)
Subjects: LCSH: Refugees. | Emigration and immigration. | Emigration and immigration--Political aspects.
Classification: LCC JV6346 .M34 2023 | DDC 325/.21--dc23/eng/20230508

1 2 3 4 5 28 27 26 25 24

Designed by Jon W. Stoy

Table of Contents

Acknowledgments

We, as editors, would like to take this opportunity to thank the numerous individuals who helped make this project possible. Of course, this manuscript would not have been a reality without the assistance and guidance of the AUC Press, particularly the help of Anne Routon. In the early stages of the project, thanks must be given to the Center for Migration and Refugee Studies (CMRS) at The American University in Cairo (AUC), especially Director Dr. Ibrahim Awad, Dr. Maysa Ayoub, and all the staff who supported the project in a myriad of ways; the AUC Centennial Scholarship Fund and the Office of Research, Innovation, and Creativity at AUC who funded the two-day workshop in Fall 2019 on "Researching Migration and Mobility in the Global South," which became the guiding framework for this edited volume; and the AUC HUSS Lab project, Public Humanities for Egypt and the Global South, funded by the Andrew W. Mellon Foundation, who provided incalculable support for the two-day workshop in Fall 2019 and was instrumental in the funding of the original copyediting of the manuscript.

In addition, this project would not have gotten off the ground without the support of AUC CMRS students Julia Hause, Jonathan Hearn, and Ziteife Okechukwu Obi, who assisted in the planning and logistics during the workshop. Adam Eddouss and Alaa Kasmo, AUC CMRS students, also helped us a great deal in checking the formatting of each chapter.

Abbreviations

3RP	Regional Refugee Response Plan
AKP	The Justice and Development Party
ASEAN	Association of Southeast Asian Nations
BROUK	Burmese Rohingya Organization UK
CBO	Community-based Organization
CEVA	Center for Empathy and the Visual Arts
CODHES	The Consultancy for Human Rights and Displacement
CSO	Civil Society Organization
DRC	Democratic Republic of Congo
ECOWAS	Economic Community of West African States
ELN	Ejército de Liberación Nacional/Colombian National Liberation Army
EPL	Ejército Popular de Liberación/Popular Liberation Army of Colombia
EU	European Union
EUCAP	European Union Capacity-Building Mission in Niger
EUTF	EU Emergency Trust Fund for Africa
FARC	Fuerzas Armadas Revolucionarias de Colombia/ Revolutionary Armed Forces of Colombia
FGDs	Focus Group Discussions
FRA	European Union Agency for Fundamental Rights
FRC	Free Rohingya Coalition
FRONTEX	European Agency for the Management of Operational Cooperation at the External Borders of the European Union

FSLN	Frente Sandinista de Liberación Nacional/National Sandinista Liberation Front
GDP	Gross Domestic Product
ICAO	International Civil Aviation Organization
IDP	Internally Displaced People
INEC	The National Institute of Statistics and Census of Costa Rica
INGO	International Nongovernmental Organization
IOM	International Organization for Migration
ISIS	Islamic State in Iraq and Syria
JEM	Justice and Equality Movement
LFIP	The Law on Foreigners and International Protection
M-19	Movimiento 19 de Abril/19th of April Movement
MFA	Ministry of Foreign Affairs
MINSA	Ministerio de Salud de Nicaragua/Nicaraguan Ministry of Health
MoSS	Ministry of Social Solidarity
MoU	Memorandum of Understanding
NGO	Nongovernmental Organization
OBE	Officer of the Order of the British Empire
OBMigra	Observatório das Migrações Internacionais/International Migration Observatory
OIC	Organization of Islamic Cooperation
PAC	Partido Acción Ciudadana/Citizens' Action Party
PEP	Special Permit of Permanence
RCP	Regional Consultative Process
RSD	Refugee Status Determination
SLA/M	Sudan Liberation Army/Movement
UN	United Nations
UNDESA	United Nations Department of Economic and Social Affairs
UNHCR	United Nations High Commissioner for Refugees
VAT	Value-Added Tax
WHO	World Health Organization

Foreword

Ibrahim Awad

This book is about migrations in the Global South. It brings out similarities and differences between migration in the Global South and Global North and takes up knowledge production in the field. It is the fruit of an initiative launched by the Center for Migration and Refugee Studies (CMRS) at The American University in Cairo (AUC) in 2018. The book comprises fourteen chapters written by seventeen scholars researching migration in the different regions of the Global South. The fourteen chapters are ordered in four parts on navigating knowledge regimes in various national contexts; state politics and global governance; migrants, im/mobility, and migration regimes; and nation-state and nationalism in and through migration contexts.

In a world of nation-states, population movements between its units are assumed to be generated by the same drivers, to face similar hurdles, and to produce comparable outcomes that could be addressed by a set of not-too-divergent policies. This assumption is based on the fallacious premise that nation-states are all similar political entities, rightfully entitled to preserve their national communities with the provision of respecting human rights. The fact is that modern nation-states are shaped by their history, their locations in the structure of the global economy, and their populations' composition. These factors and geography, in turn, determine the ability of each nation-state to preserve its national community and even its commitment to do so. Nation-states in Africa, Asia, and Latin America, including the Middle East, have come about in historical processes different from those of the Global North. Nation-state formation in Bolivia, for example, has little, if anything, to do with that in France. The same applies to Mali and Germany, Bangladesh and Belgium, and Italy and Jordan. In the Global South, in more cases than not, administrative demarcations drawn by colonial powers became political borders at the independence of the new nation-states. The function of administrative demarcations was to facilitate the administration of the territories by, often, the same colonial power. It was not to mark frontiers in accordance with the Westphalian logic, which then allowed to mold coherent, supposedly homogenous, national communities, as national borders are expected to do. The administrative demarcations, turned into national borders, often split ethnic groups and were obstacles to natural economic, cultural, and social processes. To be sure, the challenges of molding homogenous national communities also confronted nation-states in the Global North and were the origin of repeated human tragedies. One notable difference, though, is that in the Global North, nation-states faced these challenges during periods in which they were ascending in their quest for power and competing for it among themselves. Their national communities had also sufficiently integrated.

The international border regime aims at enabling states of the Global North to preserve their national communities and to admit in their midst only those foreign elements that they consider useful for their survival and prosperity. The Global South is supposed to make effective the same international border regime. In contrast to the Global North, however, many of its nation-states face the challenges of the inherited artificial national borders that split ethnic groups and make the integration of separate national communities difficult. In fact, the interest of these nation-states may lie in facilitating the reconstitution of the split communities and the processes they shared. As discussed in the book, from the effective operation of the border regime, the Global North also seeks to preserve its national communities from migration from the Global South. However, the nation-states in the Global South, because of their underdeveloped location in the structure of the global economy, cannot keep their populations from seeking better futures in the Global North and in other countries that share its appealing possibilities. States in the Global South are also often unable to control their entire territories, which extend to their artificial borders. Most importantly, the international border regime does not recognize the reality of demand for migration in the Global North. This unrecognized demand often makes the efforts that the Global North states exert to control their borders futile.

There is controversy about the existence of an international migration regime. Most scholars deny the existence of one. Recognized and accepted international norms and rules may be questioned. Yet, it can still be argued that at least a de facto global migration regime exists, constituted by the set of norms, regulations, and practices at the national and regional levels. These are complemented by existing migration rules at the international level that are not universally recognized. In line with regime complexity, international migration is also governed by norms of human rights, refugees, and international trade. The interpretation of the complex regime differs, with nation-states selecting from among its norms or distorting them at implementation. In discussing irregular migrant readmission and border externalization, the book discusses this selection and distortion.

The border regime defended by the Global North rests on a conceptualization of international migration that does not sufficiently recognize the limitations of nation-states of the Global South and minimizes its own demand for migration. The very high percentage of migrant workers in essential occupations during the pandemic in Europe reveals the importance of this minimized demand. As indicated in Carlos Sandoval-García's chapter, countries of destination in the Global South, such as Costa Rica, shared this high representation of migrant workers in essential occupations during the pandemic, which brings out similarities between South and North.

The book and CMRS do not aim at marking yet another artificial demarcation line separating the Global South from the Global North. There is one encompassing global system. In fact, despite the differences in their genesis, nation-states of the Global South and Global North share commonalities, such as the concentration of migrant workers in essential occupations. Discrimination and xenophobia against migrants are other negative shared practices they, or parts of their populations, resort to at times. The book and CMRS wish to broaden the scope of international migration, including refugee studies. So far, the migration and refugee studies discipline was developed to study population movements between and toward states of the Global North, their impact, and the policies required to address them. To further its studies, the discipline produced concepts adapted to their subject. A good number of scholars of the Global North extended their interests to population movements in the Global South. Some studied migration, especially refugee movements, to the Global South. But their interest in population movements from and in the Global South was mostly examined from the perspective of migration to the Global North. This is perfectly justified. Research is undertaken about the concerns of the community in which the researcher is situated. This adds tangibility to the research. However, left at that, research in international migration leaves the discipline amputated from large processes of population movements. Deprived of closely examining international migration in the Global South, its processes, actors, governance, and outcomes on their own merits, the discipline is incomplete. By launching the initiative and with this book, CMRS aims to contribute to expanding the field of the discipline and to enrich it.

But the effort exerted in the book goes even further. It examines practices at the global level and scrutinizes established concepts and their widespread understanding. Is an individual either a migrant or a refugee? Can they not be both at the same time? The conflation of concepts thus arises as an issue that should have consequences. In this particular case, for instance, the conflation would have consequences for international law.

Thus, this book aims to produce knowledge that should result in the long-overdue expansion of the international migration discipline. The research community and, beyond it, migrants, refugees, nation-states, and the regions in which they are situated should be the ultimate beneficiaries of the effort encapsulated in the book.

I wish to thank all the authors for their efforts in producing the following solidly argued chapters. My appreciation also goes to the four editors who ensured the coherence of the book and the smooth transition between its parts and chapters. I finally express my gratitude to my colleagues Gerda Heck and Elena Habersky, who managed the entire process besides their role as editors, from the launching conference held at the American University in Cairo in 2019 until the book in your hand was published.

Introduction

Gerda Heck, Eda Sevinin, Elena Habersky, and Carlos Sandoval-García

This edited volume came into being as part of a workshop organized by the Center for Migration and Refugee Studies (CMRS) at The American University in Cairo in October 2019. Colleagues from different institutions in Africa, Asia, and Latin America and various disciplines, including anthropology, cultural studies, ethnomusicology, international relations, media studies, political science, and sociology, contributed to the workshop and the book with their analyses of migration patterns, forms of displacement, migration trajectories, legislations, and migrant agency. Discussions at the workshop helped us point out several common issues and delineate peculiarities that prevent both "provincial" views and crude generalizations. Learning from different regions was perhaps the main outcome of the workshop and what inspired this collection. The workshop and the edited volume that followed confirmed how scarce the academic exchanges are between higher education institutions located in different regions of the Global South; clearly, a pending task largely postponed. In short, the workshop and the edition of this collection aim to follow and analyze the transnational connections, mobility, and the politics of migration in the Global South.

Throughout the workshop and the editing process of this volume, we discussed certain concepts circulating in the migration and refugee studies field. A significant part of these concepts has arguably been developed in and against the background of the Global North context. Throughout the entire process, we wanted to put these concepts and theories to the test in order to expand and stretch them with particular attention to different contexts in the Global South. In doing so, we hope to initiate a conceptual discussion in migration and refugee studies regarding not only how migration and mobility unfold in the Global South but also whether and how the dominant theories and conceptualizations developed so far in migration scholarship enable or obscure our understanding of mobilities in the Global South.

Our point of departure is twofold: On the one hand, data show that international migration increased from 2 percent to 3.5 percent between the first decades of the twentieth and twenty-first centuries, and this proportion has tripled since 1970 (IOM 2019, 20). However, this increase is not nearly as high as the pace of politicization of migration, as 96.5 percent of the world population has not engaged in a particular form of cross-border mobility that is considered migrating. Nonetheless, migration has been framed as a "problem," a "crisis" that needs to be managed, sometimes resorting to emergency measures. Many electoral campaigns, predominantly in the Global North but also in the Global South, include migration as a topic of debate, which is frequently framed in border security and controls. Predominantly, yet not exclusively, right and far-right political organizations and parties usually articulate their discourses around halting migration. Particularly after the so-called "migration crisis" in 2015 in the European Union, North America, and Australia, border securitization once again became a prevailing topic.

A consequence of this politicization of migration seems to leave the impression that most migration takes place from South to North and East to West, neglecting many other forms of mobility in other regions. However, data at the global scale provide a different picture than the dominant understanding of migratory flows. In 2017, about 90 million people migrated between the countries of the Global South, whereas the number of people who migrated from the Global South to the Global North was around 85 million (UNDESA 2017). At the same time, 84 percent of the world refugee population, estimated at 20.4 million, was residing in Africa and Asia (Awad and Natarajan 2018). The data we provided above is collected by international organizations such as IOM, UNDESA, and UNHCR, which rely on certain categories of mobility while ignoring (if not obscuring) the global asymmetries of mobility, especially the unsanctioned mobilities (see also Castles 2012). The figures might be higher if these estimations considered undocumented and unauthorized cross-border movements, especially between African countries. In addition, given that the borders and boundaries between the Global South and Global North are geographically vague and indeterminate and culturally, politically, and economically permeable, it would not be wrong to claim that the breadth and volume of mobility within the Global South are even higher.

However, especially following the summer of 2015, which was dubbed the "long summer of migration" (Kasparek and Speer 2015), discussions around human mobility were framed as a "problem" or a "crisis" that locates the West/North at the center of political geography, and that needs to be solved—through emergency measures, no less—by keeping refugees and migrants away from the borders of Northern countries. Such discussions reproduce the conception that migratory movements are necessarily unidirectional, starting in the Global South with a telos to arrive at the desired destination, the Global North. Challenging this conception requires much more careful attention to how migration policies, governance, social and economic relations, forms of knowledge production, and the arts evolving around migration have been shaped in other parts of the world. In addition to contemporary context-sensitive analyses, challenging dominant representations of mobilities also requires historicizing current border regimes and relations evolving around mobility. In her contribution to this volume, Danyel Ferrari shows how, through representations of refugeehood in public artworks, the Global North is depicted and reenacted as the ideal destination for refuge. She particularly analyzes how grand-scale public artworks focus on refugee deaths in transit work as "awareness-raising projects" to reaffirm the value of Europe both as an ideal place to be and as a space of humanitarianism. Reversing the gaze from the Southern representations in the North depicted by Ferrari to the Northern presence in the South, Sally Souraya contributes to this volume with her photo essay depicting a refugee camp for Syrians in Lebanon. Through showcasing donors' logos and other foreign expressions on the flimsy tents, Souraya, through her images, depicts the lasting power and influence of the Global North within the Global South when it comes to the policy of containment of refugees close to where they originated and far away from the Global North (Chimni 1998).

Certainly, such representations are not limited to the contemporary world. For Chimni (1998, 359), the politics of mobility unfold as "the contemporary struggle for global space." This is by no means a new phenomenon with long-lasting causes rooted in globalized capitalism, colonialism, and imperialism. In fact, diplomatic, military, economic, and political relations between the Global South and the Global North around the topic of migration are central pillars "in the maintenance of our postcolonial present" (Walia 2021, 42). Allison Wolf's chapter, focusing on Venezuelan emigration to Colombia, the world's most prominent contemporary exodus—from a country with the largest proven oil reserves globally to Colombia, a traditional country of emigration—highlights the importance of locating forced migration and displacement within larger historical processes. In a similar vein, in their contribution to this volume, Tasneem Siddiqui and C R Abrar focus on the case of Rohingya refugees of Myanmar and show how the colonial past is still very relevant in the state, granting people status as well as non-status. In other words, they keep track of the historical (colonial and postcolonial) events that have made the Rohingya first Burmese citizens, then temporary residents, and finally refugees in Myanmar's neighboring countries, as well as show how such struggles have been affected by the geopolitics revolving around Northern and Southern global powers.

Focusing on migration and mobility in the Global South, we do not aim to reinforce South/North or East/West binaries based on marked differences between the geographies or to reproduce the political, economic, social, and cultural ways that demarcate the face of the earth into nation-states as well as into regions bearing power asymmetries. Regardless of region, many nation-states are responsible for grave human rights violations in and through migration and asylum regimes, as well as dispossession and forced displacement of millions of people. Countries of the Global South, as much as the Global North, have brought migration and mobility to the forefront of political discussions, mostly to the detriment of migrant and refugee populations. Nonetheless, instead of focusing on differences between the imagined geographies of North and South, we suggest that the way the discussions have unfolded in the Global South should be addressed in their contexts, yet in a relational manner.

Migration politics and policies in the South show a variety of narratives that do not follow a single pattern. For instance, in Turkey, welcoming refugees has been made a part of populist discourse, and violent border regimes were concealed behind this populist rhetoric without necessarily making it a central discussion for the electoral campaign. In this volume, Eda Sevinin elaborates on how, in Turkey's current policy, humanitarianism and "crisis discourse" are used to legitimize and justify Turkey's migration and asylum management, which is rather based on a humanitarian government of refugees instead of fully recognized rights and status. Hence, she argues that the humanitarian production of the refugee allows for biopolitical interventions: in Turkey's case, make stay and let go. Jordan was also seen as very welcoming to Syrian refugees during the early stages of the onset of the crisis in 2011. In 2013, Jordan closed its border with Syria, citing security concerns. However, it has been argued that the country was attempting to show the supposed magnitude of the influx to the international community in order to receive a large sum of financial assistance (Tsourapas 2019), which is also why it turned to encampment for many Syrians (Turner 2015).

On the other hand, imageries of South-to-South migration in Latin America show that although derogatory ways are not just mimicry of North narratives, these are no less xenophobic. Terms such as "*bolitas,*" "*venecas,*" or "*nicas*" to name Bolivians, Venezuelans, and Nicaraguans in Argentina, Colombia, or Costa Rica, respectively, illustrate ways in which the other—usually but not exclusively the migrant—is indispensable for representing nation identity and nationhood (see Wolf and Sandoval-García in this collection). In turn, these imaginaries are often translated into immigration legislation, around which there is a continuous dispute between views that emphasize migrants' fundamental rights and authoritarian perspectives that promote nativist policies. These examples show that when looking at the dynamics of migration, it is pivotal to take the particularities of different regions into account.

In the editing process of this book, the entire world faced the outbreak of the COVID-19 pandemic in early 2020, impacting all walks of life as well as the landscapes of migration and border politics. Complex inequalities in economic conditions, mobility, and healthcare, but more importantly in accessing resources, became more visible and striking with the pandemic. At the beginning of the pandemic, borders quickly closed to non-nationals due to safety measures put in place by national governments and airline companies. Labor migrants found themselves placed in camp-like settings in countries of destination as fears of the virus spreading from foreigners skyrocketed at the beginning of the pandemic (Pattison and Sedhai 2020). In addition, refugees waiting to be resettled were left waiting in countries of asylum as resettlement numbers plummeted to the lowest numbers in decades (UNHCR 2021). Xenophobic and anti-migrant politics took a new appearance by pointing to refugee and migrant communities as dangerous intruders of the "national body." As Nicholas De Genova (2021b, 286) aptly puts it, "with the rising panic of the COVID-19 coronavirus pandemic, this perceived menace of migration predictably came to be reframed as a contagion of suspect, unruly, unwashed bodies, presumptive carriers of infectious diseases and ghoulish viruses."

After two years of living with/in the pandemic, borders are gradually opening to "legitimate" passengers with no sign of allowing further access to migrants and refugees or improving the lives of migrants by the host governments. Nonetheless, mobility continues, and people strive to undo the border regimes even amid an unprecedented public healthcare and economic crisis. In his chapter, Carlos Sandoval-García reflects on the impacts of COVID-19 and the immediate closing of borders in Costa Rica on Nicaraguan migrants increasingly facing resurging xenophobia. In doing so, Sandoval-García powerfully points to a paradox that became starker during the pandemic: Nicaraguan migrants who were, at the time of the pandemic, labeled as the potential carriers of disease were also undocumented labor migrants whose labor power was indispensable for the Costa Rican agriculture. This paradox, however, has not been translated into improving the well-being of migrants in Costa Rica or other parts of the world. Instead, exposing the commonalities between the poor "nationals" and migrant/refugee communities in other parts of the world, migrants and refugees crossing borders, as well as impoverished people, were "disproportionately relegated to conditions of precarity, abandonment, and expulsion" (De Genova 2021a, 241).

Mutual Interdependencies of Border Regimes in the Global North and Global South

Although we particularly focus on migratory movements in the Global South, we are well aware of the need to emphasize interdependencies between the migratory regimes of the Global South and the Global North. This interdependency is coined by unequal relationships and historically determined power hierarchies and asymmetries between different parts of the world. Border externalization processes are, among others, the most concrete manifestations of these interdependencies. Since the end of the 1990s, almost simultaneously, border externalization has emerged as a central policy framework for the European States, the USA, and Australia, attempting to control and manage migration movements long before they reach the physical borders of the Global North.

By border externalization, we, drawing on Casas-Cortes, Cobarrubias, and Pickles (2015, 895), mean "a fundamental change in the scales and operations of border institutions." It is widely used to account for the strategies of Global North countries to geographical, scalar, and administrative displacement and relocation (but also multiplication) of borders and border management in the countries of the Global South. The European Union's Neighborhood Policy, the agreements between the United States and Mexico, and Australia's decision to confine migrants in a number of Pacific islands are probably the most known externalization of border policies (Frelick, Kysel, and Podkul 2016). As Thomas Faist (2019) aptly notes, the main aim of such policies by the Global North was "to offload the costs of control to countries of origin and transit" but also to bail themselves out of the human rights violations that have been on the rise at the borders as well as in the migration regimes of the Global North since the 1970s.

Using the terminology "border regime," we refer to debates in migration and border studies, in which "border" is seen as deterritorialized and more or less pervasive rather than a line surrounding national territories. "Border regime analysis" allows one to analyze the border as a space of conflict and contestation between various actors trying to govern the border and migration movements. This analysis includes national governments, international and national organizations working with migrants and refugees, as well as NGOs, community-based organizations, and the migrants and refugees themselves, which are contesting, day by day, border restrictions and border securitization (Hess, Kasparek, and Schwertl 2018; Karakayalı and Tsianos 2007; Casas-Cortes et al. 2015).

Border externalization policies, an indispensable part of contemporary border regimes, also mirror power asymmetries embedded in and constitutive of global politics. Consequences of this externalization politics are depicted in Leander Kandilige and Thomas Yeboah's chapter in this volume, discussing how European migration politics toward West Africa are undermining and hindering the ECOWAS free movement protocols from 1979, which seek to promote intraregional mobility and socioeconomic development in West Africa. But externalization politics are not limited to nation-states. It includes international organizations such as IOM that are operative in different parts of the world and collaborate with governments, local organizations, and communities. The assemblage of various actors with different capacities and aims shows that border externalization policies and practices go beyond interstate relations from North to South well into globalized capitalism by curiously merging military-industrial complexes such as the European Union's Frontex with the humanitarian sector, including local and international NGOs.

In this edited volume, we suggest complicating the border externalization policies in two respects: firstly, we highlight that although increasingly harsh and violent, border regimes often fail to stop migratory movements of people on the move, as do border externalization policies. Migratory movements have often been depicted as objects that are steered at borders, but not as a (co-)constitutive factor of the social and political transformation of societies and of borders themselves (Federico and Hess 2021). People continue their mobilities in the face of various obstacles, and many of them embark upon "fragmented journeys" (Collyer 2010) either by way of deportations and comebacks or through waiting in transit contexts. Elena Habersky shows in her chapter how the Darfuris, through the desire to escape a war-torn homeland, would experience a protracted fragmented journey, taking them from Darfur to Amman and, for a large segment, back to Khartoum and onward to Cairo. These forms of mobility, in addition to transnational ties produced by these journeys between countries, regions, and, at times, continents, go beyond physical mobility and open a space for transnational negotiation in terms of ideological, religious, and political ideas. Evrim Hikmet Öğüt's article in this volume is also an example of such journeys. Discussing how Chaldean-Iraqis in Turkey's transit context experience temporality beyond past and present by adding another aspect—the future to be built elsewhere—Öğüt shows that the Chaldean-Iraqi community in Istanbul negotiates geography, temporality, and transnationality through musical practices cutting across various languages and geographies from their hometown to Turkey and the Global Northern countries of resettlement such as Canada and the US.

Therefore, such nation-state–centric policies emerge not as "solutions" to the increasingly securitized movement of humans but as "a more complex institutional and legal geography [that] is being constructed to both channel and combat the diverse flows and forces of migrants" (Casas-Cortes, Cobarrubias, and Pickles 2015, 909). Furthermore, in response to new tactics and strategies of border externalization, people always find and make new routes to carry on their mobility. Along with other examples featured in this volume, Mélanie Montinard analyzes the "*woutes*" (routes in Haitian Creole) followed by Haitians who leave Brazil in an effort to make it to the USA. Roads taken by the Haitians, more complicated—if not perilous—in time, illustrate the extent to which migration controls oblige migrants to take longer routes yet are not able to hinder mobility altogether.

Secondly, contributions in this volume highlight that the Global South countries are not passive implementers of migration policies imposed upon them by the countries of the Global North. They shape, challenge, disregard, and at times reverse these policies based on and in accordance with their nation-state policies. As mentioned above, border externalization policies are also widely used by Global South countries through various diplomatic, military, and economic means. Moreover, the way it has become ubiquitous is based on global power hierarchies and is not only related to the power asymmetry between the Global North and the Global South but also within the Global South. David Bolaños Acuña's photo essay appears as a testimony to these complicated power hierarchies by showing irregular Nicaraguan sugarcane workers, who risk life and limb, working in sun-drenched Costa Rican fields for hours on end as they barely make enough money to get by.

Knowledge Production in Migration Studies

Another issue that is discussed by the contributors of this edited volume is knowledge production in migration studies. It has been well established in scholarship that public discourse and academic research are highly concentrated on seemingly linear and unidirectional border crossings from Global South countries to Global North countries (South-to-North migration). This being the case, conceptual frameworks and theories developed in and with regard to Northern contexts are sometimes uncritically applied to other contexts, which, in turn, obscure local (and indigenous) knowledge, methodologies, and theories. *Making Routes* aims to critically approach the current discussions and broaden this landscape by offering analyses of South-to-South migration in Africa, Asia, and Latin America, seeking to show continuities and contrasts since the dialogue and comparison between cases in these continents are not encountered often in the scholarship. For example, in her essay, Sara Sadek analyzes how geopolitics and donor interests are impacting knowledge production by refugee actors and local scholars in the case of Egypt. She shows how research agendas, including design and implementation in Egypt, are heavily impacted by the dynamics of the global refugee regimes, including confining the focus of research to socioeconomic analysis to enhance avenues for refugees in Egypt to keep them from reaching Europe as much as possible.

We further want to draw attention to two potentially problematic issues that have been quite dominant in migration scholarship by not only academics but also international organizations, NGOs, and community-based organizations. The first one is that the geographical demarcation between the Global North and the Global South goes well beyond neutral geographical lines. Especially after the decades-long Cold War rivalry between East and West was replaced with a purportedly more ideologically neutral South–North dichotomy, the Global North came to represent certain values such as the free market, representative democracy, and individual liberties (Sheppard and Nagar 2004). In contrast, the Global South has been widely used to denote certain countries and regions that lack these qualities or that are "yet to come." Therefore, the Global South came to be associated with poverty, underdevelopment, authoritarian regimes that suppress individual rights and liberties, and susceptibility to internal conflicts and civil wars (Fiddian-Qasmiyeh 2020). This being the case, Global South countries are readily associated with "migrant- and refugee-producing" contexts as opposed to the Global North as the "desirable destination countries."

Nonetheless, the lines between the two are rather vague. The construction of the Global North and the Global South is contingent upon globalized capitalism and neoliberalism, as well as "a network of political and economic elites spanning privileged localities across the globe" (Sheppard and Nagar 2004, 558). That is, as Fiddian-Qasmiyeh (2020, 8) points out, the South and the North—historically, politically, and economically—are constructed and "'evolve in a mutually constitutive relationship' rather than in isolation from one another."

It is not only difficult but also epistemologically and methodologically problematic to assign reified characteristics to differently located yet highly interdependent and interwoven regions of the world. The Global North and the Global South are intertwined in such complex ways that indigenous communities, people of color, migrants and refugees, and impoverished populations within the geographical boundaries of the so-called Global North countries have more in common with populations of the Global South than the populations of the Global North as typically conceived and represented. Based on these observations, scholars have argued that "there are multiple 'Souths' (and Southern voices) within powerful metropoles, as well as multiple Souths within multiple peripheries" (Fiddian-Qasmiyeh 2020, 8).

Despite the multiplicity, interdependency, and permeability, the separation and contrast between the Global South and the Global North prevail in the migration and forced migration scholarship, somehow retrenching the global hierarchies between the South and the North. Chimni (1998) argues that as early as 1983, but particularly after the end of the Cold War, the differences between the so-called First World and the Third World countries were being forcefully asserted. The so-called differences, which Chimni (1998) calls "the myth of difference," are attributed to and justified by "stressing the differences between . . . the historical and political contexts" of these regions. Concentration on reified differences between the regions was later projected onto the global regime of mobility that is highly hierarchical and uneven.

Secondly, human mobility encompasses a huge diversity of patterns, motivations, and modalities everywhere, as much as in the Global South. It has become increasingly visible and effective in conditioning the lives of mobile people from the Global South in creating various legal and political categories that aim at locating people on the move within constructed yet powerful and effective (if not life-determining) statuses such as refugee, migrant, asylum seeker, and internally displaced person. Although seemingly less related, we believe that this aspect is also an integral part of knowledge production practices in the sense that people's mobility is confined in various conditions, sets of rather limited yet largely changing rights, and entitlements based on epistemic and legal taxonomies. These categorizations and taxonomies bear the aim of ordering people's mobility based on state-centric understandings, the roots of which can be traced back to the colonial periods where the orderly and desirable forms of mobility were accorded to white middle-class men. As shown in Gerda Heck's chapter on the migration and mobility strategies of Congolese believers, one can see how migrants can be traders and refugees at the same time. They can shift from being a migrant to being a refugee only if the legal system allows this, or become a trader if this supports their mobility.

Other local and indigenous forms that challenge dominant forms of mobility, which have been practiced for decades, if not centuries, are disclosed and marked as "illegitimate" and "disorderly" and are eventually criminalized and/or rendered deportable. Dominant asymmetries of movement inscribed in the global hierarchies between and within regions that have their roots in the colonial past are still very much relevant in the knowledge production practices of migration scholarship in our postcolonial present. Contributions featured in this volume have also delved into the complexities of partaking in global academic knowledge production of migration scholarship. Duduzile S. Ndlovu and Kudakwashe Vanyoro, drawing on their respective ethnographic research, focus on the tension between the need to negotiate researchers' position within the global academy to produce knowledge that is globally relevant while ethically responding to local concerns and social codes. This being the case, instead of insisting on attempting to—sometimes forcefully—fit people on the move in various categories, contributions in this volume recognize this diversity and complexity of mobility and people's right to refuse forced mobilization and provide on-site explorations of various contexts.

Structure of the Volume

The first section of the volume, "Navigating Knowledge Regimes in Various National Contexts," focuses on reflexive accounts of researchers regarding knowledge production and representation of refugees in various contexts. Chapters in this section range from the importance of researchers' reflexivity and positionality in exploring migration in the Global South to artistic representations of refugees and their impacts on how migration researchers, activists, artists, and migrants approach and conceptualize various mobility experiences in the Global South.

The section starts with Sara Sadek's chapter. Based on ethnographic research in Egypt between 2013 and 2017, Sadek's contribution in this volume focuses on how the global refugee regime and geopolitics affect the production of knowledge on refugee communities by local researchers and practitioners. Sadek's chapter emphasizes the importance of reflexivity and positionality of a local researcher in order to shed light on how migration researchers in the Global South contexts navigate their position within global power relations and local dynamics.

The next chapter further delves into the politics of migration research and the location of researchers in the African continent. Drawing upon case studies of research with Zimbabwean migrants in South Africa, Duduzile S. Ndlovu and Kudakwashe Vanyoro describe how African migration scholars based in the continent have to negotiate their position within the global academy to produce knowledge that is globally relevant while ethically responding to local concerns and "social codes." African migration scholars, they argue, are often tasked with "data collection," while Global North partners frame the research questions and "build capacity," depending on the donor's priority. This "extractive" relationship creates ethical tensions for African migration scholars working within "rigged" neoliberal academic spaces that perennially raise the urgent need to "decolonize" the university.

In the final chapter of this section, Danyel Ferrari changes the direction to the art world and unsettles the call for "empathy" and "awareness-raising" projects that have burgeoned in internationally well-known artworks as the dominant representations that relay experiences of migration and mobility. Ferrari looks at how, since the spring of 2016, numerous public artworks appeared in Europe that addressed forced migration and the increasing deaths of migrants in transit as they crossed the Mediterranean. The most visible of these were created by internationally recognized artists, who were decidedly outside of the refugee communities they represent and were widely disseminated via social media content production sites and were configured, by both the artists and those sharing them online, as "awareness-raising" actions and calls for public "empathy" for migrants. Her chapter explores how the dominant representations of migration in several of the most visible of these projects reinforce hegemonic narratives of South-to-North/East-to-West trajectories and, specifically, the desirability of European destinations in their construction of geographies of empathy.

The second section of the book, titled "State Politics and Global Governance," opens with Sally Souraya's vivid ethnographic photo essay. Souraya's photographic contribution looks at global migration through the (literal) lens of a refugee camp in Lebanon. Souraya's photography, echoing Paul Virilio's well-known conceptualization of "integral accident," shows how the very tangible folding of the tarps of the tents (reading "Made in Germany" in one and "Border" in another) in a refugee camp exposes the presence, if not embeddedness, of the Global North's border externalization in the Global South. The section continues with academic contributions with a shared focus on how national and global politics shape, produce, and reproduce forms of differential inclusion and exclusion as well as mobility and immobility.

In her contribution, Eda Sevinin discusses the political implications of how Turkey used its "migrant/refugee crisis" discourse to (re-)locate itself in global geopolitics. Looking at both media and political rhetoric, she argues that the "migrant/refugee crisis" was attributed to the failure of the Western/European countries to live up to the promises of certain ideals—humanity, human rights, and conscience—to which they subscribe. With this rhetoric, the Turkish nation-state, including the media and parts of civil society, postulated a moral primacy over the Western imaginary of human rights.

The next chapter calls attention to the importance of historicizing experiences of mobility and immobility by showing how the Rohingya, through nation- and state-building processes, have been transformed into refugees in a region where they have lived for centuries. Tasneem Siddiqui and C R Abrar's chapter provides a complex historical-structuralist analysis by combining the historical with the contemporary. They focus on the Rohingya by examining not only the impacts of colonization and decolonization processes and the history of Myanmar's nation- and state-building but also the geopolitics and pursuit of interest by powerful states of the Global South and the Global North. Authors, however, go beyond existing discussions and make heard both the Rohingya and the Burmese dissident groups in the historical and contemporary debates and refer to their demands for justice, accountability, and return of the Rohingya to their country of origin with dignity.

Finally, Leander Kandilige and Thomas Yeboah refocus discussions on mobility in a regional context explaining how global mobility regimes and border externalization policies engender new borders and boundaries. Kandilige and Yeboah critically examine how regional integration attempts in West Africa are both aided and frustrated by the externalized migration management interests of the European Union. Drawing on several empirical cases and examples from the research literature, they argue that in their attempt to restrict movement to Europe, European migration policies toward Africa have further contributed to criminalization of migration, experiences of horrific sufferings by migrants, and more importantly, restriction of migration within the African context, thereby undermining the goal of free movement protocols that seek to promote intraregional mobility and socioeconomic development in West Africa.

As mentioned above, in the face of progressively vicious border regimes and further impediments on mobility, people continue their mobilities. Migrants resist and subvert restrictive immigration and border policies while confronting increasingly violent and, at times, lethal obstacles daily. Chapters in the third section, titled "Migrants, Im/Mobility, and Migration Regimes," address the dynamic ways in which migrant and refugee communities *make routes* and embark upon various, often fragmented, journeys to and from different localities, most of the time subverting and undermining fortified borders and racialized and exclusionary immigration politics.

Mélanie Montinard explores the experiences and dynamics of Haitian mobility on the various *woutes* ("routes" in Haitian Creole) that Haitian migrants take from Brazil to pursue *lavi miyò* (a better life). By taking the perspective of people on the move, she describes the movement of individuals and families in response to the immigration laws enacted by various governments and immigration authorities in the Americas. Delineating the strategies that are developed to cross borders, she shows how these strategies are deeply rooted in the networks and cultural practices and how the mobility impacts not only the ones who are migrating but also the ones who remain.

Elena Habersky's contribution to the volume focuses on the fragmented journeys undergone by the Darfuris who fled violence in Sudan and face other forms of violence in their place of destination, in this case, Amman and Cairo. Although confronted with violent border regimes, the Darfuri refugees resist and extend the national borders in their everyday life, which is not intrinsically tied to Darfur, Cairo, or Amman. Through her reflective ethnographic account, Habersky shows that their relationships, hopes for the future, solidarity, and struggle to be situated go well beyond the particular locations in which they live.

Gerda Heck follows the journeys of migrants from the Democratic Republic of the Congo as they try to forge new lives in the cosmopolitan cities of Istanbul, Rio de Janeiro, and Guangzhou. Tracing increased migration of the Congolese since the 1980s, Heck explores a diasporic religious landscape that has arisen globally. By conducting field research in Congolese revival parishes that have been established, she looks at how much churches have impacted and transformed the physical and social landscapes in these three cities due, in part, to Congolese migration. Her chapter closely examines the entanglement of transnational mobility, immobility, personal aspirations, religion, and trade.

Evrim Hikmet Öğüt explores the lives of Chaldean migrants, the most populous Christian community in Iraq, who utilize Turkey as a transit country. First, exploring migration governance in Turkey and the history of Iraqi-Chaldean migration to the country, Öğüt looks at the apparent differentiations of migrants who use a country as transit in the production of music rather than a believed final destination. Music, she argues, is ground for understanding the migratory experience, thus, the specific circumstances and awareness of impermanence in transit migration. Thus, she exemplifies some basic aspects of the Chaldean-Iraqi migrant community's musical practices in Istanbul, focusing on the specific characteristics of the transit migratory experience.

Finally, the section focusing on "Nation-State and Nationalism in and through Migratory Context" opens with David Bolaños Acuña's photo essay, which highlights complicated power hierarchies in the Global South. His photography powerfully depicts the life and struggles of sugarcane field workers in Guanacaste, Costa Rica, who happen to be irregular migrants from Nicaragua. Besides working in perilous conditions, Nicaraguan sugarcane field workers also face a life-threatening kidney disease caused by working in the sun for long hours. Acuña strikingly explains, despite such risks, "sugarcane harvest, or 'zafra,' is their way of earning their livelihoods but also possibly losing their lives."

The following two chapters by Carlos Sandoval-García and Allison Wolf, respectively, extend the investigations in this book by bringing in the analysis of how, in the context of migration, nationalism erupts and becomes the hegemonic filter for the larger sections of the host societies. Their chapters show how nationalism and nation-state centrism in addressing migration are interwoven with economic, political, and social inequalities.

Sandoval-García's chapter explores the forced displacement of Nicaraguans into Costa Rica, particularly as a result of the deep political crisis arising from the 2018 protests against Daniel Ortega's government and in the context of the COVID-19 pandemic. This article ends by reflecting on the links between the closing of borders by the Costa Rican government due to the spread of COVID-19, the refusal of the Nicaraguan government to adopt physical distancing measures, and the xenophobic response by sectors of Costa Rican society.

Allison Wolf addresses international Venezuelan immigration, estimated at nearly 17 percent of the total population (28.5 million), which is considered the world's largest exodus by the Organization of American States (OAS). About 30 percent of Venezuelan migrants are established in neighboring Colombia. This situation was a relatively sudden reversal of circumstance since Colombians were forced to leave their country as a consequence of the most prolonged military conflict in Latin America. Despite the fact that about 10 percent of the Colombian population currently lives abroad, Venezuelans in Colombia face several forms of discrimination and xenophobia.

References

Awad, Ibrahim, and Usha Natarajan. 2018. "Migration Myths and the Global South." *The Cairo Review of Global Affairs*, no. 30: 46–55. https://www.thecairoreview.com/essays/migration-myths-and-the-global-south/.

Casas-Cortes, Maribel, Sebastian Cobarrubias, and John Pickles. 2015. "Riding Routes and Itinerant Borders: Autonomy of Migration and Border Externalization." *Antipode* 47, no. 4: 894–914. https://doi.org/10.1111/anti.12148.

Casas-Cortes, Maribel, Sebastian Cobarrubias, Nicholas De Genova, Glenda Garelli, Giorgio Grappi, Charles Heller, and Sabine Hess. 2015. "New Keywords: Migration and Borders." *Cultural Studies* 29, no. 1: 55–87. https://doi.org/10.1080/09502386.2014.891630.

Castles, Stephen. 2012. "Methodology and Methods: Conceptual Issues." In *African Migrations Research: Innovative Methods and Methodologies*, edited by Mohamed Berriane and Hein de Haas, 31–70. Trenton, NJ: Africa World Press.

Chimni, B. S. 1998. "The Geopolitics of Refugee Studies: A View from the South." *Journal of Refugee Studies* 11, no. 4: 350–74. https://doi.org/10.1093/jrs/11.4.350-a.

Collyer, Michael. 2010. "Stranded Migrants and the Fragmented Journey." *Journal of Refugee Studies* 23, no. 3: 273–293. https://doi.org/10.1093/jrs/feq026.

De Genova, Nicholas. 2021a. "Life versus Capital: The Covid-19 Pandemic and the Politics of Life." *Cultural Dynamics* 33, no. 3: 238–45. https://doi.org/10.1177/09213740211014335.

———. 2021b. "On Standby . . . at the Borders of 'Europe.'" *Ephemera* 21, no. 1: 283–300. https://ephemerajournal.org/sites/default/files/2022-01/21-1De%2520Genova.pdf.

Faist, Thomas. 2019. "Contested Externalisation: Responses to Global Inequalities." *Comparative Migration Studies* 7, no. 45: 1–8. https://doi.org/10.1186/s40878-019-0158-y.

Federico, Veronica, and Sabine Hess. 2021. "Protection Regimes: A Critical Analysis." *Global Migration: Consequences and Responses*—RESPOND Working Paper Series. http://urn.kb.se/resolve?urn=urn:nbn:se:uu:diva-438808.

Fiddian-Qasmiyeh, Elena. 2020. "Introduction: Recentering the South in Studies of Migration." *Migration and Society: Advances in Research* 3, no. 1: 1–18. https://doi.org/10.3167/arms.2020.030102.

Frelick, Bill, Ian M. Kysel, and Jennifer Podkul. 2016. "The Impact of Externalization of Migration Controls on the Rights of Asylum Seekers and Other Migrants." *Journal of Migration and Human Security* 4, no. 4: 190–220. https://doi.org/10.1177/233150241600400402.

Hess, Sabine, Bernd Kasparek, and Maria Schwertl. 2018. "Regime ist nicht Regime ist nicht Regime. Zum theoriepolitischen Einsatz der ethnografischen Grenz-Regimeanalyse." In *Was ist ein Migrationsregime?* edited by Andreas Pott, Christoph Rass, and Frank Wolff, 257–83. Wiesbaden: Springer.

IOM. 2019. *World Migration Report 2020*, edited by Marie McAuliffe and Binod Khadria. Geneva. https://publications.iom.int/system/files/pdf/wmr_2020.pdf.

Karakayalı, Serhat, and Vassilis Tsianos. 2007. "Movements that Matter." In *Turbulente Ränder: Neue Perspektiven auf Migration an den Rändern Europas*, edited by Transit Migration Forschungsgruppe, 7–22. Bielefeld: Transcript Verlag.

Kasparek, Bernd, and Marc Speer. 2015. "Of Hope. Ungarn und der lange Sommer der Migration." Bordermonitoring EU. https://bordermonitoring.eu/ungarn/2015/09/of-hope/.

Pattison, Pete, and Roshan Sedhai. 2020. "Covid-19 Lockdown Turns Qatar's Largest Migrant Camp into 'Virtual Prison.'" *The Guardian*, March 20, 2020. https://www.theguardian.com/global-development/2020/mar/20/covid-19-lockdown-turns-qatars-largest-migrant-camp-into-virtual-prison.

Sheppard, Eric, and Richa Nagar. 2004. "From East-West to North-South." *Antipode* 36, no. 4: 557–63. https://doi.org/10.1111/j.1467-8330.2004.00433.x.

Tsourapas, Gerasimos. 2019. "The Syrian Refugee Crisis and Foreign Policy Decision-Making in Jordan, Lebanon, and Turkey." *Journal of Global Security Studies* 4, no. 4: 464–81. https://doi.org/10.1093/jogss/ogz016.

Turner, Lewis. 2015. "Explaining the (Non-) Encampment of Syrian Refugees: Security, Class, and the Labour Market in Lebanon and Jordan." *Mediterranean Politics* 20, no. 3: 386–404. https://doi.org/10.1080/13629395.2015.1078125.

UNDESA. 2017, *Population Facts*. https://www.un.org/en/development/desa/population/publications/pdf/popfacts/PopFacts_2017-5.pdf.

UNHCR. 2021. "With Refugee Resettlement at a Record Low in 2020, UNHCR Calls on States to Offer Places and Save Lives." https://www.unhcr.org/news/news-releases/refugee-resettlement-record-low-2020-unhcr-calls-states-offer-places-and-save.

Walia, Harsha. 2021. *Border and Rule: Global Migration, Capitalism, and the Rise of Racist Nationalism*. Chicago: Haymarket Books.

PART ONE

NAVIGATING KNOWLEDGE REGIMES IN VARIOUS NATIONAL CONTEXTS

1

On the Threshold: Conducting Research about Refugees in Egypt

Sara Sadek

Introduction

This chapter engages with wider debates on the impact of geopolitics on the role of actors: local researchers and practitioners in the refugee field of Egypt. The focus is the less visible interlocutors of knowledge who contribute to public discourses on refugees in countries of asylum. It focuses on the author's experience as an actor in the refugee regime, emphasizing the "reflexivity" and "positionality" of a local researcher in Egypt. The chapter engages with a number of key questions: How is knowledge produced by refugee actors, including local scholars, affected by geopolitics, such as donor interests? What implications does this have on the conduct and outcomes of knowledge produced about refugees in Egypt? The chapter's findings are based on ethnographic research conducted in Egypt between 2013 and 2017[1] for a doctoral thesis concerning the role of "civil society organizations" (CSOs)[2] as advocates for refugees in Egypt.

The Refugee Regime

There is a close link between debates about displacement and state sovereignty. Global power relations and negotiations between states do not only affect the operations of the relevant actors, but they also constitute a key factor in knowledge production on refugees in countries of asylum. The years after 2011 have witnessed the emergence of top global priorities to discuss the impact of events in the Middle East and North Africa (MENA) and notably the "Syrian Refugee Crisis." New concepts have become very common in global and regional fora, such as "burden-sharing," "migration governance," and "securitization of migration." Such concepts reflect the operations of local refugee regimes, including in Egypt.

The Global Landscape

According to various scholars, the contemporary global refugee regime has been marked by the revival of similar challenges and problems to that of the early 1950s (Betts and Loescher 2014). Such issues include the fragmentation and tensions between humanitarianism and state interests and preferences, lack of appropriate durable solutions, and inadequate refugee protection and asylum systems. Based on the 1951 Convention on the Status of Refugees upheld by the United Nations High Commissioner for Refugees (UNHCR), the global refugee regime is based on two main foundations: burden-sharing and asylum (Betts 2010, 16). This seems to overlook that "forced human migration" involved fewer numbers in premodern history and was not bound by international cooperation (Haddad 2008, 62). Additionally, negotiations around burden-sharing covered fewer refugees and were also confined to what is currently referred to as the Global North. In his pioneering piece, Chimni (2000) highlights what he referred to as "the birth of a discipline" of forced migration and refugee studies. Chimni's argument is that such studies and the rise of humanitarianism have emerged to fulfill hegemonic powers' interests and to label the unwanted, displaced populations from the Global South (Chimni 2000).

Since 2015, and with significant numbers of Syrians fleeing the conflict in their homeland, which started in 2011, the notion of burden-sharing has been indirectly replaced by negotiations to "manage migration." This was particularly obvious with the "Syrian Crisis" eruption described by the concerned European Union (EU) countries. After being neglected at earlier stages of the conflict, initiatives to address migration management have operated in a complex manner on the national, regional, and international levels (Kalm 2010, 21). Globally, *The New York Declaration for Refugees and Migrants* of 2016, adopted by the UNHCR, served to enhance cooperation to protect refugees and vulnerable migrants as well as expand funding to promote protection. Subsequently, *The Global Compacts for Refugees* were commissioned. Similarly, *The Global Compact for Safe, Regular, and Orderly Migration* aims to serve similar outcomes to the refugee compact but with a broader target group. In the abovementioned discussion of relations around migration and states, there seems to be an assumption that the leading players in global migration policy are the hegemonic states in the Global North, which influence the various regimes reflected in UN agencies and other players such as UN agencies, INGOs, and NGOs, who in turn influence those states hosting asylum seekers and refugees. There is minimum consideration given to the role played by local actors, either as UNHCR's partners or grantees from embassies and funding agencies. Analysis of CSOs has been limited to looking at their scope of work from the perspective of being the implementing and operational partners of UNHCR (Betts 2010; Gatrell 2016). The analysis is usually focused on understanding the operations of UNHCR rather than analyzing the work of such partners. Such CSOs, especially in Egypt, have a significant role to play not only in responding to policies but also in reshaping knowledge and discourses around refugees.

Betts and Loescher (2011) analyze the refugee regime as a complex regime in which the UNHCR is no longer the only UN actor, as was the case upon its creation in 1951. Instead, there is competition between new emerging UN organizations. Such transformations, according to them, affect not only the operations of UNHCR but also the receiving states as they may strengthen or challenge paradigms of protection (Betts and Loescher 2011, 69). Betts and Loescher have visually depicted the various overlapping regimes, including human rights, labor migration, development, travel, security, and humanitarian regimes, plus their organizations and areas of intersection with the refugee regime (Betts and Loescher 2011, 73). They highlight that the most problematic overlap and contradiction are between the refugee regime and the travel regime under which organizations such as the International Organization for Migration (IOM), Regional Consultative Processes (RCPs), International Civil Aviation Organization (ICAO), and EU mobility partnerships operate. The contradictions are due to the nature of the travel regime that is based on the foundations of state-to-state negotiations to restrict and govern migration, hence, according to Betts, blocking channels of asylum (Betts and Loescher 2011, 74). In addition to UN actors, the availability of several non-state actors has given states the privilege to address any specific issue concerning their interests by selecting the organization that would cater best to its needs (Betts and Loescher 2011, 70).

The Refugee Regime in Egypt

On the policy level, Egypt is a signatory to the 1951 Convention on the Status of Refugees, the 1967 Protocol, and the 1969 Convention Governing the Specific Aspects of Refugee Problems in Africa. Egypt has made reservations to five articles in the 1951 Convention: Article 12 (1) (personal status), Article 20 (rationing), Article 22 (1) (access to primary education), Article

23 (public relief and assistance), and Article 24 (labor legislation and social security). However, it has allowed access to public education for Sudanese, Syrian, and Yemeni refugees and access to primary healthcare. For the past decade, there have been several efforts to issue an asylum law in Egypt to govern issues related to refugees and asylum seekers. However, to date, it has not materialized. Alternatively, there are laws concerned with entry and exit and with combating human smuggling and trafficking in persons.

Throughout recent decades, the key actors involved in the refugee regime in Egypt have been governmental entities and the United Nations High Commissioner for Refugees as the key UN agency in cooperation with various UN agencies. For implementation, UN agencies have collaborated with various international and local nongovernmental and non-UN actors, including international nongovernmental organizations (INGOs), locally registered nongovernmental organizations (NGOs), and faith-based organizations, namely churches and community-based organizations (CBOs) run by the refugees themselves. Legally, each of the entities mentioned is governed by different legal frameworks in Egypt. INGOs have considerable flexibility in receiving funding linked to funding channels through their headquarters and regional offices. Legally, they fall under the Ministry of Foreign Affairs (MFA). Local NGOs fall directly under the Ministry of Social Solidarity (MoSS), governed by Law 70, enacted in 2017, and specifically listed clauses for authorized activities and procedures for foreign funding approvals by MoSS. Churches catering to the needs of refugees follow their federations. They, for a long time, have been deprived of access to foreign funding by key funding agencies in Egypt as a requirement put by the donors. Since 2017, this requirement has been eased, which has benefited churches' ability to receive funding, as do other INGOs and NGOs. CBOs remain the category facing the most challenges due to the lack of capacity to form local NGOs. The latter require Egyptian partners among their board of directors and other challenges such as financial and capacity issues. They often receive funding either through the community or through the employment of their staff as "community leaders" serving and implementing programs on behalf of INGOs, NGOs, and faith-based organizations.

Waves of refugees fleeing to Egypt have shaped the emergence of actors serving refugees in Egypt. Between the 1950s and 1990s, the number of refugees in Egypt under the mandate of UNHCR was quite small. Hence, the refugee regime in Egypt was not yet established in its current form. In 1954, the Egyptian government signed a Memorandum of Understanding (MoU) with the UNHCR. The following areas were delegated to the UNHCR:

protection, registration, documentation, Refugee Status Determination (RSD), and resettlement to persons of concern.[3] The influx of larger numbers of refugees in Egypt started in the 1990s (Grabska 2006; Al-Sharmani 2003), which has affected the structure of the refugee scene in Egypt. In Egypt, UNHCR has played a major role in providing services and protection to refugees under its mandate and, in many cases, through collaborating with international and local actors known as implementation or operational partners. Some of them are partially funded by the UNHCR, while other agencies fund others. Such organizations have played an essential role in complementing socioeconomic, counseling, and legal services.

Throughout recent decades, the categories of actors serving refugees have included other UN agencies, international nongovernmental, faith-based, community-based organizations, local nongovernmental organizations, and academic institutions. However, starting in the early 2000s, the number of organizations concerned with refugees or including refugees in their target groups has vastly increased and expanded. Between 2008 and 2010, a transformation took place in the legal aid and counseling services provided to refugees in Egypt. New formalized organizations were formed based on the recruitment of refugees for outreach purposes, yet owned or managed by foreign experts. In the early 2000s, most of the refugee services were legal aid ones, supporting refugees in RSD preparation; they were initially refugee-led organizations until they became more formalized.

Additionally, a few CBOs were representing different segments of the Sudanese, Ethiopian, and Somali communities. They have acted as a platform for community cohesion, agency, and support (Huser 2014). However, they have usually been very short-lived and have gradually lost support from donor agencies. The establishment of critical legal aid and community organizations between 2000 and 2009 was attributed to the presence and efforts of the late scholar, legal advocate, and Officer of the Order of the British Empire (OBE) Barbara Harrell-Bond, who significantly contributed to shaping the local refugee regime in Egypt. The formation of new entities was instigated by the influxes of Iraqi refugees but was also extended to include preexisting and new influxes of African refugees.

Although the conditions and needs of refugees in Egypt remained largely the same after 2011, the landscape of CSOs has slightly changed due to regional and national factors. In light of the political events and conflicts in neighboring countries such as Syria and Yemen, new influxes arrived, particularly from Syria. INGOs started expanding their activities into the refugee scene through newly formed "emergency" programs.[4] The new cooperation with INGOs was facilitated by some regional response frameworks, namely the Regional Refugee Response Plan (3RP) targeting Syrians in neighboring MENA countries. The additional regional impact of the changes in MENA and Sub-Saharan regions and migratory routes led to evolving protection needs and changing migrant smuggling and human trafficking patterns. The second factor reflects local political events instigating the closure of some local organizations for legal reasons. Since some refugee-concerned entities could not fulfill the requirements of registration enlisted by Law 70 governing NGOs in Egypt, their work was jeopardized, and they were eventually shut down. Zetter (1996) argues that local NGOs catering to refugees are struggling with the competition from INGOs mostly preferred by donors. In Egypt, the tension and discrepancy are based not only on the North/South CSOs divide but also on the formal/informal organizations. Funding channels by donors and UN agencies were directed toward those operating within the legal framework in Egypt, including the different regulations governing INGOs and local NGOs.

Raper (2003) looked at another dimension of such power relations in refugee aid, which is the relation between bigger INGOs and local NGOs and CBOs working with refugees and how the former "cannibalize" the latter's operations. He explains the limitations local CSOs and smaller indigenous community organizations face in partnering with donors and bigger humanitarian organizations, who only emerge in the case of a crisis. Despite major limitations, CBOs perform an integral role for refugees (Griffiths, Sigona, and Zetter 2005; Huser 2014). The majority of CBOs in Egypt lack operational and financial capacities.[5] However, they have the strongest links to less visible groups of refugees (Crisp, Morris, and Refstie 2012) and often find themselves being a source of access for larger CSOs. What exacerbates their conditions is not only a lack of capacity but also the fact they are unable to obtain proper legal recognition. This leads organizations with a more stable legal status or relations with donors to use links with CBOs as a crucial strategy for getting more donor funds.

Understanding the burden-sharing negotiations' impact on asylum countries like Egypt cannot overlook the rise of the prevailing concept of "transit." During the 1990s, the transit concept had started to emerge in relation to EU migration. The term applies to the context of Egypt but has been used increasingly for several reasons. First, due to Egypt's geographic location and perception as one of the key transit routes toward the EU, albeit outflows have significantly shrunk, the perception has continued among donor agencies. Second, due to the above reservations made to the 1951 convention and the fact that Egypt has no encampment policy, Egypt's local perception of asylum has been marked with temporariness. Third, UNHCR is considered the key entity concerned with refugees in Egypt, which has led it to become a form of "surrogate state" (Kagan 2011; Slaughter and Crisp 2009). Few scholars have contested the term (Collyer, Düvell, and de Haas 2012; Hess 2012; Collyer and de Haas 2012; Schapendonk 2012). In the last few years, the term *transit* has remained undefined but codified for irregular migration. Despite the common use of the term, its definition is blurred and not possible to identify in light of the dynamic nature of migration (Collyer and de Haas 2012). The term's conditions and appropriateness to the context have been overlooked (Collyer, Düvell, and de Haas 2012). What would be more appropriate to the context of Egypt is a concept that encompasses the characteristic indefinite transit and "continuous status of waithood" (Auyero 2011).

Producing Research about Refugees in Egypt

Within the global and national frameworks discussed in the previous sections, scholars and practitioners have found themselves in critical positions to produce knowledge on refugee issues in Egypt, unintentionally affecting discourses around refugees, including advocacy potentials for better lives for refugees in Egypt. In many cases, research and practice have unintentionally promoted an image of helplessness and constructed vulnerability rather than promoting positive discourses and creating avenues for refugees in Egypt. On the contrary, producing knowledge around refugees has been clearly shaped by geopolitical factors in the form of research agendas, funding channels, and practicalities and ethics of conducting research and needs assessments on refugees.

In Egypt, research concerned with refugees has been conducted in two key ways. The first is through academic institutions and/or visiting scholars conducting research studies on refugees in Egypt. This approach is often marked by short, intense encounters with refugee community members, whereby mutual benefit is rarely ensured. In many cases and for accessibility purposes, researchers volunteer or approach CSOs to access community members. The second approach is through assessments done by UN agencies solely or in collaboration with other partners, namely academic institutions, CSOs, and/or CBOs. The production of knowledge also happens through comparative regional assessments such as Regional Refugee Response Plans (3RP), in which the local government may be a partner. In recent years, visibility studies and needs assessments have been concerned with a small number of overlapping topics in Egypt: opening avenues for refugees and examining socioeconomic conditions with particular emphasis on (child) protection, education, health, and livelihood. Such topics are usually on the top agenda of donor agencies who strive to better the situation of refugees with less emphasis on the local sociolegal and political framework in Egypt, which could enhance refugee life prospects. This focus has also largely ignored refugees' preferences and their specific needs, which might be different from those identified by donor agencies and implemented by refugee sector actors, including UN agencies and civil society.

Research on socioeconomic issues that is determined by donor agencies and implemented by actors in the refugee regime has implications in terms of research agendas and leads to severe methodological and ethical ramifications affecting the outcome of the research. Scholars have discussed issues related to the invisibility of refugees and the hidden nature of such communities (Harrell-Bond & Voutira 2007; Krause 2017). Such difficulty in locating samples has implications on the outcome of research findings regarding the representativeness of the population at hand and, hence, their needs (Jacobsen and Landau 2003). Practitioners reiterated their inability to reach out to less visible groups due to the urban setting framework, which has resulted in knowledge deficiencies of the actual needs of refugees, exacerbating mistrust toward actors serving them, including UN agencies as well as international and national NGOs. Alternatively, they have used their caseloads as sample frames. This may lead to another key challenge in which CSOs find themselves constantly performing the role of gatekeepers of the community to formulate sample frames (Polzer 2013) that later inform services and practices catered to refugees (Krause 2017; Jacobsen and Landau 2003).[6] Krause highlights that the basic principle of "do no harm" in refugee research is often breached if research is conducted in precarious conditions (2017). He also critically analyzes the "undercover" research conducted in anthropological settings. Service providers conducting research using databases of case records and statistical information concerning their caseload to identify needs and gaps in services also raises important concerns.

Beyond the adequate sampling challenge, mistrust has significantly affected research around refugees (Rodgers 2004; Lammers 2007; Hynes 2003; Krause 2017). Hynes (2003) lists factors behind mistrust from and to refugees and different entities. Mistrust is very common toward refugee organizations generally. Many refugees decline to approach organizations because such actors only collect the information to seek more funds toward operation costs. In many cases, CSOs find themselves acting as translators and "intermediaries" for larger agendas set by donors or international organizations

(Merry 2006). Such dynamics are exacerbated in a context like Egypt. In the past few years, a few incidents have affected the relations between refugees and UN agencies, including the 2005 Mustafa Mahmoud sit-in[7] that lasted for three months and closures of UNHCR's operations twice in 2011 and 2013 for security reasons following political events in Egypt.

Other issues are important to consider, such as "giving back" to the community, which is rarely achieved, whether a UN, an academic, or a civil society organization has conducted research. Actors take part in such "needs assessments" and, in some cases, disseminate the research through events and public reports and, in many cases, keep them internal for programmatic purposes or due to sensitive information or criticism of services received. Moreover, assessments led by various organizations are marked by being conducted hastily, rarely capturing the actual needs of the target groups included in the study. This is due to rushed timelines of project budgets or feasibility studies expected by donors to be conducted in a very short amount of time. As mentioned earlier, because of issues of accessibility, such sample frames are usually composed of the caseloads of the CSO acting as the research implementer for donor agencies. They often serve particular agendas of larger donors, academic institutions, and UN agencies.

Technically, few organizations have the capacity to adequately conduct such assessments. They end up focusing on the groups already included in their caseload and repeating exactly the same research. In a training[8] given to representatives of CSOs, UN agencies, and early researchers, it was observed that representatives lacked the basic ethical and methodological skills to conduct social research. As expressed by practitioners, research steps are taken unsystematically without a particular order or scientific approach.

The research process conducted by academics and practitioners alike in Egypt has also been marked by very limited engagement of refugees in the research agenda. In recent years, refugee community members have been recruited to conduct fieldwork interviews and Focus Group Discussions (FGDs) (Hynes 2003), which support the philosophy of involvement of the community in designing research. But in reality, they implement a design *already* prepared by the organization. They are only required to get the sample frames, which again raises the same concerns of representativeness. The limited role of refugees themselves as knowledge producers is a key point of concern. As Jones (2015) points out, the refugee field in Egypt has been heavily dependent on foreign experts, who are constantly replaced. To give a better image of inclusion, many INGOs and local organizations alike hire refugee "leaders," "focal points," or "outreach coordinators," which seems, in principle, an adequate approach for many reasons. First, it ensures the involvement of refugees in their field, enhancing a sense of agency. Second, it leads to the desired outcome of providing a livelihood opportunity for refugees. Third, it ensures the organizations are kept up to date with the most recent trends directly from the refugee community. However, in reality, there are a lot of challenges in the approach of hiring refugees. First, they are usually extremely underpaid in comparison to international experts or Egyptian staff members. Many of them work for months and years and get paid without signing contracts, even in formal organizations, which raises critical questions of accountability concerning financial reporting (Kaldor 2003). Additionally, community leaders who work in the majority of refugee organizations usually have better chances for resettlement (Jones 2015). They are usually among the most knowledgeable about community needs, the most qualified and well trained, all reasons for organizations to want to hire them. They also endure the highest risks of being targeted, sometimes by community frustrations or by being looked upon as receiving preferential status above other refugees, without, in fact, having connections to key organizations and resources to make a living. Finally, staff in civil society organizations who are also members of refugee communities in Egypt are usually used to represent the predicaments of refugees in meetings hosted either by UN agencies or donors. In such meetings, directors who are either international or Egyptian ask to be accompanied by their "refugee leaders" or "community outreach staff." The main person in charge usually conducts presentations in a hierarchical manner. The floor is given to refugee attendees only when accurate information is needed concerning a new community trend or to reiterate certain needs by the community. They are rarely allowed to present their involvement in operations or programmatic aspects.

In terms of conducting research, such refugee focal points are often resorted to form sample frames and prepare FGDs, but their role in shaping the actual research findings remains quite limited. Based on an account of a representative of one community in Egypt who also manages a local NGO:

> UN agencies and INGOs simply approached me to solely use my contacts in the community. In order for them to reach the target groups for their programs, they obtained my lists of refugees, many of them were very hard to locate residing on the outskirts of Cairo and Alexandria. I was paid a small amount as a consultant and then later on, none of these agencies wanted to formally partner with my organization. Now they are using these lists and expanding their operations.[9]

This hierarchy is not only visible between NGOs serving refugees and refugees themselves but also between organizations within the field in the form of power dynamics between bigger and smaller organizations, namely CBOs. The phrase "raising the capacity of CBOs" has been overused in budget and research proposals by CSOs with very little impact, which is reflected in the continuous dependency of such CBOs on the bigger CSOs and the lack of sustainability of many. UN agencies have preferred to work with bigger organizations as umbrellas for smaller CBOs, which has further shrunk their prospects for funding and sustainability. Some of these umbrella organizations have also been regarded by CBOs as mistrustful. Civil society organizations in Egypt serving refugees have constantly struggled with a lack of capacity, coordination, and sustainable operational plans. Yet, the discourse of lack of capacity is usually targeted toward smaller CSOs and CBOs as a strategy to maximize prospects for bigger CSOs and INGOs already obtaining a large amount of funds. There is an assumption that smaller or local CSOs cannot perform the interventions adequately, though this is often an assumption lacking clear evidence.[10] Within the same context, several organizations have unintentionally focused on what several scholars refer to as "labeling," "framing," and the concept of "performing refugeeness" in the context of refugee aid (Malkki 1996; De Voe 1981; Zetter 1996, 1991, 2007; Szczepanikova 2010; Nawyn et al. 2016). By portraying refugees as helpless and lacking agency, many organizations guarantee that they are providing justifications for donors to continue serving refugees in the same way rather than critically underscoring how refugees opt to represent themselves.

A Local Researcher on the Threshold

This section concludes the chapter by briefly reflecting the author's experience while researching with refugees in the MENA region, particularly in Egypt, since 2004. Reflexivity has been a key term in ethnographic discussions.[11] Considering the importance of reflexivity in qualitative research and specifically in ethnographic approaches, this section presents the less obvious challenges behind conducting fieldwork. It looks at the experience of being a local researcher at the threshold of being an "insider" and "outsider" in the field of forced migration in Egypt. Such positionality is usually accompanied by a "double perspective," "dual consciousness," and "double-knowledge" that create a sense of confusion (Eyben 2009; Kirby, Greaves, and Reid 2006). In approaching respondents for doctoral thesis fieldwork at a UK-based university, the author was at the same time a practitioner and consultant for different institutions, which has caused some confusion. This confusion was due to wearing multiple hats. Firstly, I worked as a researcher and later as a consultant and a gatekeeper in migration in Egypt for the past two decades. Secondly, I shifted from working purely in the academic context to working and consulting for international organizations, foreign embassies, and UN agencies. While being an actor or an insider might seem at the outset like an opportunity to have access to a wealth of knowledge and networks, it has created some obstacles during fieldwork. Such challenges were reflected in three key factors: First, in trying to avoid biases by "unlearning" some of the earlier presuppositions one had on CSOs and the refugee scene in Egypt; second, in differentiating between what was acquired in interviews for pure research purposes through observations, involvement in meetings, and consultancies with various agencies; and third, in working toward regaining the trust of actors on some issues by reiterating confidentiality and commitment not to share any information with any entity. The blurred roles and multiple strong links to various networks of refugee actors have led some respondents to be less forthcoming about certain shortcomings of their organizations or tensions with some UN agencies. This mistrust has caused some challenges of access to organizations within the field. What CSOs would tell a graduate student seeking answers to questions on challenges facing them in their dealings has been deemed different from what the same student would know in different capacities, such as informal networks and consultancies.

These fluid roles have been described by Eyben as being on the threshold between fields and roles (2009, 89):

> There are ethical (or rather moral) issues related to the possibility of wearing two hats, that of a critical anthropologist and that of a paid consultant or commissioned researcher. On the one hand, the consultancy may be undertaken in bad faith, simply as a means of access to ethnographic data without any commitment to the employer. On the other hand, if the anthropologist undertakes the commission in good faith, seriously seeking to improve the organisation's practice, then her academic peers may judge the ethnographic analysis as constrained by the researcher's instrumental engagement.

Thus, further to Eyben's (2009) stance toward wearing different hats, the author's experience has shown that membership in the field provides another layer of trust not offered to a student of migration studies as opposed to an actor in the field. Eyben called this status "a confused identity," which she experienced in social events, having access to classified stories and untold information (Eyben 2009, 98). This insider-outsider relationship was prevalent throughout years of fieldwork. This status has led to an accumulation of vast knowledge covering questions for the author's fieldwork, but much of this knowledge remained hidden in the background of analysis for ethical considerations. While the above discussion might suggest that such status has provided a good opportunity for knowledge access, in some cases, it was the reason for blocking access. Such an approach challenges researchers who not only conduct research at home but continue working or living in their field rather than performing the usual "exit" from the field as with the classical form of foreign ethnography (Eyben 2009; Kirby et al. 2006). This certainly adds to the moral obligations of being politically sensitive. In this case, the concern was to ensure that the information discussed during the research would not affect rapport and future dynamics with other actors.

In addition to the dynamics above, a knowledge seeker faced a few dilemmas as a knowledge producer on refugees in Egypt. As an actor in the refugee field, issues included adopting and implementing research agendas that are inapplicable to the local context or may not be perceived among the key priority areas. The author was faced with the dilemma of bigger organizations preferring to work with like-minded consultants who were less critical of any apparent shortcomings. In many cases, the author realized she was using a different approach based on her knowledge of the context of refugees in Egypt, being affiliated with international academic institutions and UN agencies serving particular agendas (Nagar 2014; Mama 2009) while providing little to practitioners and refugees.

Conclusion

In analyzing the knowledge production process on refugees in Egypt as a country of asylum, the chapter addresses essential predicaments related to hidden interlocutors, namely practitioners and local researchers. Donor agendas shape the design and implementation of research agendas. They have been heavily influenced by the dynamics of the global refugee regime dictating priorities to the local refugee regime in Egypt. This has confined the research focus to socioeconomic analysis to promote alternative avenues for refugees other than reaching Europe. Bound by funding challenges and the need to continue their operations, refugee actors, including practitioners and researchers, have had little input into directing such research agendas to ensure they adequately represent the priorities of refugee communities. The research approaches used with refugees in Egypt by other refugee sector actors have led to further moves to include methodological and ethical issues affecting representation, including portrayals of refugees and their supposed needs. Within this debate, this chapter has discussed the author's experience as a local researcher on the threshold due to having blurred roles as an insider and outsider, leading to tensions in implementing international research agendas that may be deemed inapplicable to the local context. The consequences are ineffective and provide only short-term engagement, with little focus on refugee rights advocacy in Egypt. The images and conditions of refugees portrayed by researchers and practitioners have often led to short-term emergency responses with less prospect for integration into national systems, fitting more the agendas of both refugee sector actors who have stayed away from advocacy. Ironically, while emergency approaches have helped agencies fulfill donor implementation for projects, they have not enhanced the potential for refugees in the long run through adequate dialogue with the national state.

Notes

1 That study used qualitative research methods, including semistructured and unstructured interviews as well as ethnographic observations (Kirby, Greaves, and Reid 2006, 134; Braun and Clarke 2013, 78; DeWalt and DeWalt 2002, 122). The fieldwork covered a duration of twenty-one months between January 2015 and October 2017, covering a total number of sixty interviews, including forty-seven interviews covering eighteen local CSOs. The study followed a longitudinal approach (Menard 1991, 5). Observations were collected during interviews and by involvement in the Egyptian forced migration field through different capacities.

2 This chapter uses the term *civil society* in the broader sense regardless of the legal status of an entity to ensure that all relevant actors concerned with refugees in Egypt are included, such as international nongovernmental organizations, local nongovernmental organizations, faith-based organizations performing similar activities to those of the previous entities, and community-based organizations. Debates on the definitions of civil society are beyond the scope of this chapter. Hereinafter, CSOs will be used in reference to Civil Society Organizations.

3 UNHCR Egypt, "What We Do." https://www.unhcr.org/eg/what-we-do.

4 Including but not limited to Save the Children International, PLAN, CARE, and Terre des Hommes.

5 Interview conducted with a representative of an organization in Egypt on September 11, 2017.

6 This finding was confirmed through two interviews conducted with representatives of two organizations in Egypt on May 14, 2017, and another interview conducted on October 3, 2017, with two organizations in Egypt.

7 For more information on the Mustafa Mahmoud sit-in, please consult: Azzam, Fateh. 2006. "A Tragedy of Failures and False Expectations: Report on the Events Surrounding the Three-month Sit-in and Forced Removal of Sudanese Refugees in Cairo, September–December 2005." Forced Migration and Refugee Studies

Program, The American University in Cairo (AUC), Cairo, Egypt.

8 The author delivered this research training to organizations in 2015 and 2016.

9 Interview conducted with a representative of a refugee-led organization in Egypt on December 3, 2015.

10 Interview with a representative of an organization in Egypt on October 17, 2017.

11 Various scholars have attempted to define it. Reason (1994) defines reflexivity as the researcher's capacity to understand his/her position and to locate it within the same social space where research is taking place. Personal reflexivity is described as "self-awareness and awareness of the dynamics between the respondents and the researcher in a fieldwork context" (Giddens 1993; Lamb and Huttlinger 1989). Reflexivity explains how the researcher's background and experience affect the different stages of the research (Koch and Harrington 1998; O'Reilly 2009).

References

Auyero, Javier. 2011. "Patients of the State: An Ethnographic Account of Poor People's Waiting." *Latin American Research Review* 46 (1): 5–29.

Betts, Alexander. 2010. "The Refugee Regime Complex." *Refugee Survey Quarterly* 29 (1): 12–37.

Betts, Alexander, and Gil Loescher. 2011. *Refugees in International Relations*. New York: Oxford University Press.

———. 2014. "Introduction: Continuity and Change in Global Refugee Policy." *Refugee Survey Quarterly* 33 (1): 1–7.

Betts, Alexander, Gil Loescher, and James Milner. 2012. *UNHCR: The Politics and Practice of Refugee Protection*. London: Routledge.

Braun, Virginia, and Victoria Clarke. 2013. *Successful Qualitative Research: A Practical Guide for Beginners*. London: Sage Publications.

Chimni, Bhupinder S. 2000. "The Birth of a 'Discipline': From Refugee to Forced Migration Studies." *Journal of Refugee Studies* 22 (1): 11–29.

Crisp, Jeff, Tim Morris, and Hilde Refstie. 2012. "Displacement in Urban Areas: New Challenges, New Partnerships." *Disasters* 36 (1): 23–42.

Collyer, Michael, and Hein de Haas. 2012. "Developing Dynamic Categorisations of Transit Migration." *Population, Space and Place* 18 (4): 468–81.

Collyer, Michael, Franck Düvell, and Hein de Haas. 2012. "Critical Approaches to Transit Migration." *Population, Space and Place* 18, no. 4: 407–14.

De Voe, Dorsh Marie. 1981. "Framing Refugees as Clients." *The International Migration Review* 15 (1/2): 88–94.

DeWalt, Kathleen M., and Billie R. DeWalt. 2002. *Participant Observation: A Guide for Fieldworkers*. California: AltaMira Press.

Eyben, Rosalind. 2009. "Hovering on the Threshold: Challenges and Opportunities for Critical and Reflexive Ethnographic Research in Support of International Aid Practice." In *Ethnographic Practice and Public Aid: Methods and Meanings in Development Cooperation*, edited by Sten Hagberg and Charlotta Widmark, 71–98. Uppsala: Uppsala University.

Gatrell, Peter. 2016. "Refugees—What's Wrong with History?" *Journal of Refugee Studies* 30 (2): 170–89. https://doi.org/10.1093/jrs/few013.

Giddens, A. 1993. *New Rules of Sociological Method: A Positive Critique of Interpretative Sociologies*, 2nd ed. California: Stanford University Press.
Grabska, Katarzyna. 2006. "Marginalization in Urban Spaces of the Global South: Urban Refugees in Cairo." *Journal of Refugee Studies* 19 (3): 287–307.
Griffiths, David J., Nando Sigona, and Roger Zetter. 2005. *Refugee Community Organisations and Dispersal: Networks, Resources and Social Capital*. Bristol, UK: Policy Press.
Haddad, Emma. 2008. *The Refugee in International Society: Between Sovereigns*. Vol. 106. Cambridge: Cambridge University Press.
Harrell-Bond, Barbara, and Eftihia Voutira. 2007. "In Search of 'Invisible' Actors: Barriers to Access in Refugee Research." *Journal of Refugee Studies* 20 (2): 281–98.
Hess, Sabine. 2012. "De-Naturalising Transit Migration: Theory and Methods of an Ethnographic Regime Analysis." *Population, Space and Place* 18 (4): 428–40.
Huser, Hannah. 2014. *A Tale of Two Somali Community-based Organizations in Cairo: Between Assistance, Agency and Community Formations*. Center for Migration and Refugee Studies: The American University in Cairo.
Hynes, Tricia. 2003. *The Issue of "Trust" or "Mistrust" in Research with Refugees: Choices, Caveats and Considerations for Researchers*. United Nations High Commissioner for Refugees (UNHCR). Evaluation and Policy Analysis Unit.
Jacobsen, Karen, and Loren B. Landau. 2003. "The Dual Imperative in Refugee Research: Some Methodological and Ethical Considerations in Social Science Research on Forced Migration." *Disasters* 27 (3): 185–206.
Jones, Martin. 2015. "Legal Empowerment and Refugees on the Nile: The Very Short History of Legal Empowerment and Refugee Legal Aid in Egypt." *The International Journal of Human Rights* 19 (3): 308–18.
Kagan, Michael. 2011. "'We Live in a Country of UNHCR'": The UN Surrogate State and Refugee Policy in the Middle East." *Policy Development and Evaluation Services: New Issues in Research*, Paper no. 201. https://ssrn.com/abstract=1957371.
Kaldor, Mary. 2003. "Civil Society and Accountability." *Journal of Human Development* 4 (1): 5–27.
Kalm, Sara. 2010. "Liberalizing Movements? The Political Rationality of Global Migration Management." In *The Politics of International Migration Management*, edited by Martin Geiger and Antoine Pécoud, 21–44. Basingstoke, Hampshire, New York: Palgrave Macmillan.
Kirby, Sandra, Lorraine Greaves, and Colleen Reid. 2006. *Experience Research Social Change: Methods Beyond the Mainstream*. 2nd ed. Peterborough, Ont., Orchard Park, NY: Broadview Press.
Koch, T., and A. Harrington. 1998. "Reconceptualizing Rigour: The Case for Reflexivity." *Journal of Advanced Nursing* 28: 882–90.
Krause, Ulrike. 2017. "Researching Forced Migration: Critical Reflections on Research Ethics During Fieldwork." *Refugee Studies Centre Oxford Department of International Development*, Working Paper Series No. 123.
Lamb, G.S., and K. Huttlinger. 1989. "Reflexivity in Nursing Research." *Western Journal of Nursing Research*, December 11 (6): 765–72.
Lammers, Ellen. 2007. "Researching Refugees: Preoccupations with Power and Questions of Giving." *Refugee Survey Quarterly* 26 (3): 72–81.

Malkki, Liisa H. 1996. "Speechless Emissaries: Refugees, Humanitarianism, and Dehistoricization." *Cultural Anthropology* 11 (3): 377–404.
Mama, Amina. 2009. "Challenging Patriarchal Pedagogies by Strengthening Feminist Intellectual Work in African Universities." In *Activist Scholarship: Antiracism, Feminism, and Social Change*, edited by Julia Sudbury and Margo Okazawa-Rey, 55–72. London: Routledge.
Menard, Scott William. 1991. *Longitudinal Research (Quantitative Applications in the Social Sciences).* California: Sage Publications.
Merry, Sally Engle. 2006. "Transnational Human Rights and Local Activism: Mapping the Middle." *American Anthropologist* 108 (1): 38–51.
Nagar, Richa. 2014. *Muddying the Waters: Coauthoring Feminisms across Scholarship and Activism*. Urbana, Chicago, and Springfield: University of Illinois Press.
Nawyn, Stephanie J., Nur Banu Kavakli, Tuba Demirci-Yılmaz, and Vanja Pantic Oflazoğlu. 2016. "Human Trafficking and Migration Management in the Global South." *International Journal of Sociology* 46 (3): 189–204.
O'Reilly, Karen. 2009. *Key Concepts in Ethnography*. London: Sage.
Polzer, Tara Ngwato. 2013. "Collecting Data on Migrants through Service Provider NGOs: Towards Data Use and Advocacy." *Journal of Refugee Studies* 26 (1): 144–54.
Raper, Mark. 2003. "Changing Roles of NGOs in Refugee Assistance." In *Forced Displacement: International Security, Human Vulnerability and the State*, edited by Edward Newman and Joanne van Selm, 350–66. New York, Tokyo: United Nations University Press.
Reason. Peter. 1994. "Three Approaches to Participative Inquiry." In *Handbook of Qualitative Research*, edited by Norman K. Denizen and Yvonna S. Lincoln, 324–339. California: Sage Publications.
Rodgers, Germane. 2004. "'Hanging Out' with Forced Migrants: Methodological and Ethical Challenges." *Forced Migration Review* 21: 48–49. https://www.fmreview.org/return-reintegration/rodgers.
Schapendonk, Joris. 2012. "Migrants' Im/Mobilities on Their Way to the EU: Lost in Transit?" *Tijdschrift voor Economische en Sociale Geografie* 103 (5): 577–83.
al-Sharmani, Mulki. 2003. "Livelihood and Identity Constructions of Somali Refugees in Cairo." Vol. 2. *Forced Migration and Refugee Studies Program*, The American University in Cairo.
Slaughter, Amy, and Jeff Crisp. 2009. "A Surrogate State? The Role of UNHCR in Protracted Refugee Situations." In *Protracted Refugee Situations*, edited by Gil Loescher, James Milner, Edward Newman, and Gary Troeller, 123–40. New York, Tokyo: United Nations University Press.
Szczepanikova, Alice. 2010. "Performing Refugeeness in the Czech Republic: Gendered Depoliticisation through NGO Assistance." *Gender, Place and Culture: A Journal of Feminist Geography* 17 (4): 461–77.
Zetter, Roger. 1991. "Labelling Refugees: Forming and Transforming a Bureaucratic Identity." *Journal of Refugee Studies* 4 (1): 40–62.
———. 1996. "Indigenous NGOs and Refugee Assistance." *Development in Practice* 6 (1): 37.
———. 2007. "More Labels, Fewer Refugees: Remaking the Refugee Label in an Era of Globalization." *Journal of Refugee Studies* 20 (2): 172–92.

2

Researching Migration on the African Continent: Reflecting on Power, Location, and Research Partnerships

Duduzile S. Ndlovu and Kudakwashe Vanyoro

Introduction

In this chapter, we explore the tensions of working within "rigged" neoliberal academic spaces (Hendricks 2018) in an attempt to unpack what it means to study Africa's mobile populations on their own terms. African universities are modeled after institutions of higher learning that emerged in the West (Hendricks 2018, 17). Hendricks (2018, 17) argues that African universities have also "cultivated their hierarchies, racial and gendered power relations, epistemologies, and ethnocentric constructions of what constitutes knowledge, and in which bodies and geographies it is supposedly located and enunciated." International research partnerships enact, expose, and compound the inequalities, structural constraints, and historically conditioned power relations implicit in the production of knowledge (Landau 2019a, 26). These long-standing patterns of power define culture, labor, intersubjective relations, and knowledge production well beyond the formal ending of colonial rule and reflect what Maldonado-Torres (2007) terms the "coloniality" of power (Vanyoro, Hadj-Abdou, and Dempster 2019).

As female and male Zimbabwean migration scholars, we both find ourselves having to navigate this ambiguous academic space. Focusing on the researcher–participant relationship, we use case studies of our research with Zimbabwean migrants in South Africa and the use of poetry, as well as socially and morally grounded reflexivity, to show what engaging in research that speaks to local concerns and interests entails. In concert, we reflect on Ndlovu's research focusing on Zimbabwean migrants' commemoration and memorialization of Gukurahundi violence in Johannesburg and Vanyoro's research exploring the experiences of Zimbabwean migrants waiting in transit shelters at the Zimbabwe–South Africa border. While this research is not part of any specific global research partnership, these dynamics are the backdrop to our work as early-career researchers seeking to gain a foothold in the academy. Furthermore, this work is located within our position as local researchers working within an institution that has been nominated as a Southern partner on several occasions. Together with colleagues of various racial backgrounds, ethnicities, nationalities, and genders, we are constantly reflecting on what our backgrounds and field presence signify for our participants.

We argue that the broader institutional context of African migration research has far-reaching implications for the researcher–participant relationship because it privileges "us" vis-à-vis the often-vulnerable migrants we encounter in our work. We propose that the use of poetry, as well as socially and morally grounded reflexivity, can present opportunities to engage in research that speaks to local concerns and interests. In this chapter, we contribute to the literature on the unequal power relationships that local or indigenous researchers have to navigate in research partnerships of inequitable power relations between the Global North and Global South and in the context of the coloniality of knowledge production within the broader academy and migration research. We reveal an extra layer of power negotiation between participant and researcher that is usually invisible to Western partners largely because they are mediated by cultural mores (Kalinga 2019). We also grapple with the challenges of our identities as researchers and their intersections with our racial, national, ethnic, and gendered identities as we experience them in the research field. The ways we experienced these fluid and situated identities (Hall 1990; Erasmus 2017) in the research field demanded attentiveness to relational ethics that we had not considered as we prepared for the research.

The structure of the chapter is as follows. We begin with an overview of current debates on research funding and collaborations between Northern and Southern scholars undergirded by questions on agenda setting and the coloniality of knowledge production. We then discuss how these power dynamics in interaction with social codes lead to some ethical challenges that emerge in particular research contexts. This is followed by a focus on our own individual negotiations of power occupying the insider/outsider position in migration research and a discussion of how this connects to the broader, ongoing debates on global research partnerships and collaborations.

Power, Location, and Research Partnerships: The Nature of Funding, Collaboration, Language, and Policymaking

The question of how migration research is framed globally, by whom, and for whose benefit has never been more pertinent. African scholars operate within a context of limited funding, resources, and precarious labor. With the proliferation of migration research partnerships through increased Western and donor interest in Africa and the Middle East, African-based migration research is increasingly becoming policy or activist-oriented (Landau 2019a). Western partners often task African scholars with managing data collection while they focus on framing the research questions, methods of inquiry, and "building capacity" (Jayawardane 2019). Of course, this all depends on the donors' or funders' priorities. Some partnerships may provide scope for African migration scholars to be part of framing the research questions. However, most collaborations are guided by a hidden, overarching agenda to do research that is meant to inform European Union (EU) policies on migration, particularly in the wake of the "European migration crisis." These policies reflect the moral panic that engulfed Europe in 2015, which led to concerted efforts to halt migration from the African continent by labeling all Africans as potential migrants capable of threatening European and African sovereignty and security (Landau 2019b).

Indeed, power imbalances can be acute in these kinds of research collaborations that span geographic and economic divides (Landau 2019a). Knowledge remains territorial and imperial as scholars produce knowledge that is always shaped and formed in a context of power (Mignolo and Tlostanova 2006). By making sense of the world in a certain prescribed way, African migration scholars become part of these power arrangements. While collaborative research may indeed lead to the development of evidence-based policies that can contribute to the alleviation of human suffering, these power arrangements often have a negative impact on African migrants for whom such work is purportedly meant. Scholars like Linda Tuhiwai Smith (1999, 1) thus suggest that the word research itself is probably "one of the dirtiest words in the indigenous world's vocabulary." Through a process of "epistemic coloniality" (Mbembe 2016), African migration researchers are intertwined in a system of knowing about others yet never fully acknowledging them. Scholars like Schinkel (2018) locate this dilemma's foundation in theorists like Durkheim, Parsons, and Habermas, which inherently means that migration research comes from a position of power—since we conduct research on "the other" (Vanyoro, Hadj-Abdou, and Dempster 2019). These theorists reinforced this power relation as they generally portray the construct of difference as an anomie or an undesirable aspect of modern societies that must be eliminated through the "necessary" processes of social cohesion, organic solidarity, and normative integration.

Some have criticized migrant integration research for studying the inability of "the other" to conform to society, not the ability of society to adapt and accept "the other" (Hadj-Abdou 2019; Schinkel 2018; Vanyoro, Hadj-Abdou, and Dempster 2019). Like other forms of "indigenous research," this kind of research takes place in a context where researchers implement concepts on the assumption of "the inferiority or devilish intentions of the Other" (Mignolo and Tlostanova 2006, 206).

Migrant integration research in many European societies continues to justify oppression and exploitation as well as the eradication of difference. In Western Europe, this occurs "amidst a public discourse that is highly toxic" (Schinkel 2018, 1). Schinkel (2018, 1) locates this toxicity in certain racisms that are hard to avoid when researchers wish to conduct research in the established way, such as acquiring research funds and publishing in high-ranking journals about immigrant integration. This immigrant integration approach problematizes the culture of the other or individuals that are racialized in particular ways by referring to society as an "unscathed whole" (Schinkel 2018). The effect of this is that society lays blame for social problems and ills on racialized individuals who reside outside of it and need to be integrated.

We argue that the uneven status quo of research funding that characterizes global partnerships, which tends to flow from Global North to Global South, has sustained the connections between this racist conception of migrants and African migration research. Local institutions generally have far less input as to how such partnerships are mediated and carried out (Jayawardane 2019). Data to frame migration prevention interventions are often generated by frequently relaying Southern perspectives to Northern policymakers and scholars (Landau 2019a). This confirms the assertion that knowledge about indigenous peoples is "collected, classified and then represented in various ways back to the West, and then, through the eyes of the West, back to those who have been colonized" (Smith 1999, 1–2). Northern partners often speak on behalf of Southern actors whose language is "too fragmented and particularistic to be globally legible" (Landau 2019a, 28). In falling back on globally legible language, migration researchers based in the West risk replicating and perpetuating established international migration narratives instead of challenging them (Pécoud 2015; Wee, Vanyoro, and Jinnah 2018). These legible ways of speaking generate "neat, orderly, and well-demarcated policy categories" out of the "messy, overlapping, and complicated" realities of migration (Wee, Vanyoro, and Jinnah 2018, 799). This limits the ways in which Southern-based researchers can shift their practice to accommodate the local concerns expressed by participants.

Activist-oriented research, however, tries to unsettle some of these dynamics. For example, participatory methods are a good way to generate evidence and engage with policy stakeholders at various stages of the research process when doing research with marginalized groups such as migrants. Participatory methods are more inclusive and utilize more iterative, flexible, visual, and captivating elements in the generation and communication of evidence. This allows for the research message to be more accessible as opposed to merely becoming policy briefs in front of policymakers that are packaged in language and categories that do not resonate with the marginal experiences of local participants. Such participatory methods may include film, photography, music, poetry, and art in order to captivate audiences and allow research to "speak for itself" in a less technocratic manner (Vanyoro 2015). Participatory methods "challenge generic definitions of evidence by providing an extension to what we call evidence," and this can be anything from a photograph to a poem (Vanyoro 2015). This approach works best when dealing with underexplored groups and issues like sex work, gay rights, and migration (Oliveira and Vearey 2015).

Research into the suffering of "others" can only be justified if alleviating that suffering is an explicit objective (Jacobsen and Landau 2003, 186; Landau 2019a; Turton 1996, 96). The dual imperative to satisfy the demands of academic peers and to ensure that the knowledge and understanding this work generates are used to protect refugees and influence institutions like governments and the UN also works against this stated objective (Jacobsen and Landau 2003, 186). In South Africa, for example, international migration narratives used to share research with policymakers can easily be used improperly to undermine the cause of protecting migrants because they align with the salience of categories of practice and migration management that rationalize the two-gate aspirations of the apartheid-colonial era Aliens Control Act. The "two-gate" policy consisted of a "front gate" for desirable white immigrants not perceived by the state as a threat to European culture, while the "back gate" was kept open to "undesirable" African migrants only because they were a cheap source of labor in South Africa's farms and mines. Throughout the 1990s, the Aliens Control Act of 1991 was the cornerstone of South Africa's immigration policy, a policy that still seeks to relegate low-skilled migrants to the "back gate." The immediate post-1994 immigration policies retained colonial "two-gate" qualities as they benefited highly skilled migrants at the expense of their less skilled counterparts (Peberdy 1998; Peberdy and Crush 1998; Moyo 2016). The researcher, in this instance, for example, may be constrained from using the local classification categories that do not fit neatly within official "globally legible" categories that justify these preexisting policies and discourses.

Globally as well, research on the relationship between migration and development tends to be framed as mutually beneficial and iteratively reinforcing. This commodification of movement leads to the exclusion of migrants whose mobility is perceived as "less developmental" in differing contexts. For example, migration does not always promote development in the neoclassical manner; it is articulated and measured by both sending and receiving states, while states also attach greater value to labor migrants over asylum seekers and refugees because they see the latter two as an economic burden. Therefore, the typical migrant addressed in global and national migration frameworks in relation to development is a labor migrant, and this suits the economic approach to migration favored by states while eschewing more difficult questions about migrants beyond their contribution to the labor markets, which would entail engaging with the question of their rights (see Wee, Vanyoro, and Jinnah 2018). To sum up, research partnerships and the ensuing "globally legible" language they generate "reproduce 'eye-wateringly' unequal power relations that basically look too much like colonial-era relations" (Jayawardane 2019, 279). In this chapter, we intervene by focusing on the interface between the researcher and the research participants as an added layer in research engagement that is rarely discussed yet essential in order to shift this dynamic.

Location and the Ethics and Social Codes of Local Research

Uneven global research partnerships also highlight the politics of using local researchers. The hiring of local research assistants and working with partner organizations in the field to help with interviews, translation, and identification of research subjects is now a widespread practice that adds to this status quo (Jacobsen and Landau 2003). Western researchers work with local researchers because they believe that working with local researchers yields better results (Jacobsen and Landau 2003). This is a form of "methodological capital" in the form of embodied capital and passing strategies (Lundström 2010, 70). While it is necessary to reach specific groups, it may also reproduce structural privileges by not engaging in difficult conversations about nationality, class, gender, and so forth (Gallagher 2000; Lundström 2010, 70).

Nonetheless, in global partnerships, expectations and ethics are not sufficiently particularized to be sensitive to such contextual realities, social codes, and dangers (Kalinga 2019). Dube, Mhlongo, and Ngulube (2014, 202) have also observed that research ethic norms are not necessarily commensurate with "indigenous moral values." Ethical norms based on confidentiality, anonymity, and beneficence may not adequately capture the morals and ethical norms of indigenous societies and may instead be disempowering (Chilisa and Ntseane 2010, 629). For example, participants may not want to remain anonymous because of the value they place on their views (Ndlovu 2010, 26–37). Therefore, it has been the case for many years now that encouraging Southern partners to collect and relay "local knowledge" incentivizes deep and myopic local engagement (Jacobsen and Landau 2003). Using research assistants from the same country or area as the respondent "risks transgressing political, social or economic fault-lines of which the researcher may not be aware" (Jacobsen and Landau 2003, 193). Writing in a similar context, Kalinga (2019, 270) observes that indigenous researchers also have "an additional obligation to respect social customs and codes," which are not easily visible to foreign research partners, and are responsible for receiving and interpreting these codes. The dilemma for African scholars is that choosing to reset the process, build trust, and address the sources of such discontent is also tantamount to "career suicide" (Kalinga 2019). The indigenous researcher thus finds themselves negotiating their place within donor priorities and, simultaneously, the expectations of the research participants.

Reflecting on the Invisible Power Negotiations between Researchers and Participants in the Field

Thus far, we have provided a broad scope of the unequal power relationships that local or indigenous researchers have to navigate in the context of inequitable power relations between the Global North and Global South and the coloniality of knowledge production. Pedri-Spade (2016, 390) asks how an indigenous person who is also a researcher within the Western academe can do "good work." This section discusses the ways we have sought to do good work as guided by the social norms around issues of class, age, ethnicity, and gender that produce certain expectations of us from participants as Zimbabweans studying Zimbabweans. Here, we do not want to assume a simplistic divide between the indigenous and nonindigenous. Mekgwe (2006, 12) explicitly argues for a focus on "the contemporary" against the trap of defining Africa in contrast to the West. Instead, our discussions are informed by our experiences in the field, which are neither necessarily of indigenous nor nonindigenous orientation but are perhaps best described as an "in-between" of protean researcher identities. These experiences revealed a gap in the ways we had framed ethical engagement within our research at the beginning of the process, leading us to question ourselves while also negotiating our early-career status and the outputs that would be rewarding toward a career in the academy. The following sections are written from the respective first-person experiences of each author.

Zimbabwean Migrants' Commemoration and Memorialization of Gukurahundi Violence in Johannesburg

In my case (Ndlovu), having a shared history of the Gukurahundi was instrumental to my gaining entry into the research site. Having a claim to Gukurahundi victimhood, as well as being married to a man of the correct ethnic identity, served to locate me as a worthy listener to people's narratives (Phoenix 2008, 81). My gendered location as a married woman mediated my access to the research and further prescribed the ways I engaged with participants and what I could and could not do or say within the research interview space.

Gukurahundi violence occurred soon after Zimbabwe gained independence in 1980. Twenty thousand people were killed, and some disappeared at the hands of the national army. The Zimbabwean government has never officially acknowledged this period (Dorman et al. 2000). The victims of the violence were supporters of the main opposition party at the time, as well as being predominately from one ethnic group. Some understand the conflict as ethnic cleansing, while others see it as primarily about politics. Many view it as a conflation of both ethnic and political differences. Participants in my study located themselves along different positions on this continuum, and many assumed in our research encounters that I held the same convictions as they did. My positionality as a woman necessitated that I do not challenge these assumptions in order to have an audience with the participants. I chose not to express contrasting views or correct participants' assumptions to be viewed as a respectful and not difficult woman, although in writing I did not have to assume the same stance. Nonetheless, I encountered resistance to the research as participants referred to similar prior encounters with "data collection." Some of these previous interactions had resulted in the misrepresentation of their stories with little to no acknowledgment of the participants. This resistance to the "harvesting" of their stories/experiences in interviews where there is no feedback loop or a clear demonstration of how their contributions link back to their lives should be read as a way in which participants sought to exercise power over the research relationships.

The recognition of my location as a member of the group thus functioned both to grant me access as well as to discipline and influence my engagement.

As an insider, I was expected to understand the ways "we" wanted "our" story to be told and to have the same demands as the participants. Participants critiqued the ways they had seen their narratives represented in simplistic ways that lacked nuance, as is often done when Global South contexts are simplified to be legible for Northern and policy audiences. They also critiqued my location within the university as one that could easily lead me to reframe their narratives in simplistic terms while also suggesting that as an "insider" they expected me to know better by telling a more nuanced story about their lives.

Having gained access predicated on my insider-outsider status, I continued my engagement with participants, directed by this culturally framed relationship. I shall present two examples where I experienced some dissonance related to age and gender. I interviewed an older man in his office. I arrived early for the interview and was told he would be arriving soon, so I waited. On his arrival for the interview, the participant proceeded to talk about what he wanted out of the interview despite my having initiated the meeting. In this interaction, I made the choice of engaging in the same way I would have had it been my father. I did not interject or interrupt the monologue about what I should be exploring in my research and how I should be doing this. He continued for a further hour in this "monologue lecture," which he concluded by delineating the things he was willing to talk to me about. I listened to the monologue and did not challenge some of the assumptions he made about my interest in the research. I did not directly challenge his views because this would violate the ways I was "cultured" and socialized to engage with a man old enough to be my father. This posture was the way I chose to engage in order to maintain my access to this interview.

The male participants' attempts at influencing the research interaction were also intertwined with the culturally defined gendered dynamics that structured it. The gendered nature of my research access invoked a questioning of to whom I was married. Here, the assumption underlying this question from the men I was interviewing was that, as a woman, I was a minor to the man I was married to. It was, therefore, important that I was married to a man of the right ethnic group. Ordinarily, this is a question I would have challenged. However, in the context of accessing the research, I had to acquiesce and respond to these challenges differently. Lykes (1997, 734) reflects on a similar conundrum where a White American female researcher in Guatemala had access to male Ladino spaces—the social gatherings of the leaders of the organizations she partnered with for her work—that Mayan women and men did not have access to. Further to this, however, she could not challenge the patriarchal jokes because this would have jeopardized her access to a space where she could learn how the organization worked. Although I had to maintain a "good daughter" or "good woman" stance within the interview settings, the writing up of the work did not require the same limitations. I had space to analyze and critique the interview transcripts in the relative safety of my research office. I could write for an academic audience without worry or care about the participants' interests, similar to the "story harvesters" participants had warned me about when they agreed to join the project.

Zimbabwean Migrant Imaginaries and Modes of Waiting at Musina on the Zimbabwe–South Africa Border

My research (Vanyoro) explored how temporal disruptions at international borders shape the meanings (im)mobile bodies attach to migration as well as the migrants' modes of waiting at the Zimbabwe–South Africa border town of Musina. Musina is the northernmost town in the Limpopo province of South Africa near the Limpopo River border with Zimbabwe (Leong 2009). The link with Zimbabwe is one of the busiest roads in the world and the busiest in Africa due to cross-border trading from Zimbabwe and people looking for employment (Leong 2009). My research documented what Zimbabwean migrants did in this in-between space, considered their objectives in staying there, how they came to terms with the potential loss or disruption of the idea of a future, and how they turned their social constructions of spatio-temporal disruptions into a resource for action. I conducted this research at two transit shelters housing men and women who were "stuck" at the Zimbabwe–South Africa border after running out of money or being robbed by human smugglers while crossing irregularly. I refer to these shelters as the Men's Shelter and Women's Shelter.

Like Ndlovu, being a Zimbabwean mediated easy access into both these spaces. Like Zaman's (2008) presence in a hospital ward in Bangladesh, my presence in Musina and the shelters "did not indicate anything unusual, unlike what would be the case of a white anthropologist in a colored population" (Zaman 2008, 150). Perhaps the most unusual thing for me was the living conditions of the men and women at the shelters. Thus, I was caught up in the complexity of researching the "other" as the "other" and navigating challenges related to my privilege and my ultimate positionality. The nature of ethnography did not prepare me to deal with questions around conducting research on poor, marginalized groups that I ostensibly belong to.

Similar to Ndlovu, my access to these two spaces was culturally defined and structured by gendered dynamics, although class also played a more specific role in my case. My gender as a male researcher meant that I was able to conduct more detailed ethnographic research at the Men's Shelter than at the Women's Shelter. The women at the Women's Shelter warily gazed at me, and being Zimbabwean meant that they would treat me as a group/community member. This membership simultaneously entailed that the social codes of dress, forms of addressing me, and gendered performances applied in our relationship. This would have made opening up to a Zimbabwean man about topics such as sex work difficult for the women because it is socially frowned upon in Zimbabwe. My male body represented the hand of patriarchy that many women are on the receiving end of back home. It also meant that I was not comfortable observing their daily routine in the intimate space of the church, where they also bathed in close proximity in this small, crowded churchyard. It was difficult to ask certain intimate questions that were pertinent to my understanding of their experiences, particularly those related to the "management" of female bodily issues like access to sanitary towels. These unspoken social codes prevented me from developing any profound or intimate relationships with these migrant women when compared to the ones I developed with the migrant men at the Men's Shelter. During four months of conducting this research in Musina, my position as a Zimbabwean "insider" allowed me to ascertain within the first week at the Women's Shelter that I would not be able to develop the kinds of relationships that would allow me to get "thick" ethnographic data. The localized notions of morality and being a good, responsible, and considerate man/woman acted as a barrier to my access.

Besides these gendered norms, there were times when other experiences challenged the assumption that I was a cultural insider by virtue of being a Zimbabwean migrant, particularly those related to class. The notion of insider-outsider is very intricate to social scientists carrying out ethnographic research and entering the field. Yet, the line between what constitutes the "inside" or "outside" in ethnographic research is often blurred (Zaman 2008). Because of my privilege, I was an insider-outsider. Sometimes people would address me in Shona, my native language, since I was visibly a black person in Musina. However, I once walked past Zimbabwean migrants and heard them whispering, "This man has the skin of people with money," despite the fact that I was dressed casually. These migrants drew on their own unspoken codes of well-being to interpret individuals' bodies in a bid to determine their socioeconomic status. These are some of the invisible issues that global partnerships hiring local researchers do not grapple with. It was also commonplace for the men at the Men's Shelter to address me in English only to reciprocate in Shona once I had initiated conversations in the dialect. Most would be pleasantly surprised, and the mantra "So you can speak Shona?" became one that I got used to, as it reflected certain presumptions they held about my identity in relation to class.

Indeed, I stayed in a hotel that had everything from laundry services to Wi-Fi. In contrast, many Zimbabwean migrants in Musina either had to return to their homes in the township or the rudimentary corrugated iron shelter in the men's case. It is, therefore, imperative, as Turner (2000) suggests, that I also involve myself in the critique of privilege: acknowledging my own privileges of class. I did not neatly qualify as an insider or an outsider within a research community; therefore, my subsequent analysis benefited from not assuming membership and making assumptions about boundaries. Like Zaman (2008), I was both an insider and an outsider who embodied both the advantages and disadvantages of this position. I was an outsider because of my class status. There is no doubt that my privilege was visible to the men from the moment I walked in each morning. Even though I did my best to tone down my clothes, it was obvious that I had woken up, taken a good hot shower, and applied lotion to myself before coming to the shelter. This toned town appearance was still at great odds with the state of the men at the Men's Shelter, who were often lacking soap or food.

This material disjuncture also gave rise to problems of reciprocity, as I would be asking poor migrant men who did not receive any meals about their hunger. This disparity quickly gave rise to dynamics reminiscent of a Dagbani proverb: "Whenever a bedridden sick person sees an old man with a white beard enter his room, he thinks a medicine man has come to cure him." Like the old man, my physical appearance generated hope that my presence would possibly solve the problems the migrants were facing. When I attended my first church service during my first week in Musina, the pastor in charge of the Men's Shelter began his sermon by asking me to come to the altar of the church. Before the church service, he had suggested that I sit in front of the church, a space that was commonly reserved for church leaders like pastors, elders, bishops, and deacons in the Pentecostal church tradition. As I stood at the altar, he introduced me by making the case to his congregation and the few migrant men in attendance that I could easily be Jesus, who was sent by God to the shelter to rescue the migrants from wasting away. He portrayed me as someone who would contribute meaningfully to the lives of the migrant men at the shelter on the condition that they showed restraint, discipline, and "respect for God." From that day, the migrant men from Zimbabwe began to perceive me in a light of moral authority. Some of those I interviewed expressed hope that when I departed with my research to wherever I came from, "we" researchers would produce the desirable results. For them, this entailed getting the Department of Home Affairs Refugee Reception Office in Musina to provide them with asylum seekers' permits, something the office had stopped doing due to the belief that all Zimbabweans were "economic migrants." My presence could easily have reinvented the same hierarchies of knowledge that I was trying to dismantle.

Responding to These Dilemmas through the Use of Poetry and Socially and Morally Grounded Reflexivity

A Note on Poetry

In my case (Ndlovu), it was imperative in the writing of the research to include the participants in the outcome of the research to make my interests as a researcher visible. Pedri-Spade (2016, 396) argues that it humanizes the "researched" when the researcher shares themselves with them. Similarly, Palmary (2009, 58) admonishes migration researchers against contributing to a system that seeks to other and police those who move. Finding ways, as researchers, to acknowledge the humanity of the researched becomes an important exercise. This was important because, in contrast to my stance within the interviews and group discussions, where I was constrained by social codes from challenging and speaking back to participants, the writing and analysis phase of the research was without constraint. As such, I wanted to give participants access to the ways I was representing our interactions. It was important for the participants to see that the gendered power dynamics that facilitated our research encounters were no longer at play. Poetry allowed my thesis to be a document written with the researched as its audience and a reimagined engagement to disrupt research's dirty history of making the researched legible for others excluding the researched (Ndlovu-Gatsheni 2017, 186).

Poetry served as a tool for reflexivity to stage my positionality and work of making meaning out of the data generated in this research. Firstly, I was able to show my researcher's location in a more nuanced way and express the subtle differences that my interactions had glossed over. Secondly, I was able to show how I was making sense of the data and what I was writing. Lastly, the poems were instrumental in summarizing the thesis, ensuring that participants not only had access to a nuanced version of my position, as opposed to my stance in the interview and my work as a researcher writing about their lives, but they could engage with the end product.[1] In this way, poetry showed the connections between my position and the research output, disputing the myth of an objective research exercise and showing that a researcher is always implicated in their work (Nadar 2014, 24).

Socially and Morally Grounded Reflexivity

Reflexivity is an important lens to engage with the social and moral codes of conducting research in context. My presence (Vanyoro) at the Men's Shelter generated expectations and hope about the possibility of me introducing changes that required political authority that I did not necessarily possess. Therefore, throughout the course of my stay, I had to correct some of these assumptions in order to get the pastor and the migrant men to understand my role and what I envisaged as possible change. I had to constantly remind the participants that I could not change the asylum system in Musina to work in their favor. Nonetheless, through sharing their stories, they could help me show "people out there" that these hidden spaces, in fact, existed so that something could be done in the future. Indeed, participating in different engagements has helped spread the word and raise awareness about these neglected and "invisible" spaces, such as the transit shelters in the border zone.

Owing to the material disjuncture that divided us, it was morally wrong for me to simply go about conducting interviews with hungry migrants. Even though traditional conceptualizations of research stress that you cannot compensate participants for their time as it compromises objectivity, I ended up reflecting on this limitation and adapting by using my own money to buy some groceries for the migrant men when I could.

In response to the gendered challenges associated with my work at the Women's Shelter, I carried out less ethnographic research there since I could not afford a female research assistant. I conducted a few interviews and scrapped the possibility of observation altogether. I conducted the interviews with the help of my partner, which neutralized the interaction and made participants comfortable to open up to another woman. As a South African, the limitation was that she does not speak Shona, so these interviews had to be conducted in English and Shona simultaneously. This was the most ethical thing to do as opposed to insisting on making each other feel socially uncomfortable. Nonetheless, in my subsequent writing up of the thesis, I felt it was best to write more extensively on the experiences of the migrant men at the Men's Shelter and not to misrepresent the experiences of the migrant women at the Women's Shelter, as the interviews were not as rich as I would have wanted.

Conclusion

Location as an insider-outsider carried the benefits of easy access to the field because we understood the language and social codes that governed interactions. This location also opened up a space for participants to negotiate the terms of our access. We endured great discomfort due to the expectations of power that participants assumed we had. These were mediated in different ways by our gendered identities. For Ndlovu, role-playing the "good" Ndebele woman gave access to the field while also constraining who she could be within this space. In Vanyoro's case, his gender constrained his interactions with potential female participants. It necessitated a different engagement, while his class status was read as bringing much-needed hope to the men's shelter. Since the exploration of the gender dynamics of waiting is ongoing and an incomplete aspect of the project, Vanyoro is now planning to hire a female Zimbabwean research assistant to capture the women's experience of waiting in Musina through thicker descriptions and elaborate ethnographic observation.

As Zimbabweans researching Zimbabwean migrants, we were familiar with the social codes and could easily assume the stance required to gain access to our field site. The social codes mediating this access facilitated the participants' resistances and subtle attempts at directing the research trajectory. Our negotiations with participants were not always commensurate with our forecasted research trajectory and necessitated a different engagement beyond what we had anticipated during the planning phase. Current research proposal development and ethics review processes do not sufficiently anticipate or prepare researchers for challenges such as the ones that we faced. As researchers located in the Global South, we were challenged. We had to negotiate our place within these dynamics together with the demands of the academy vis-à-vis those of the communities we belong to.

We began the discussion with an overview of the broader debates on research partnerships and relationships focused on the global scale, the funder and the funded, and Northern/Southern locations. Our contribution to these debates focuses on the often-overlooked dynamics at the interface of the researched vis-à-vis the local researcher, research assistant, or fieldworker. What happens within the interaction between indigenous researchers and participants is significant if we are committed to undoing the colonial legacy of research in these contexts. The "researched" are important stakeholders within the research relationship, and we need to pay attention to the ways we engage with them. Our reflections here show the limited yet important ways that participants want to influence and drive the research agenda and how poetry, as well as morally and socially grounded reflexivity, can respond to these dilemmas. This calls for deeper reflection on the role we play as "Southern" researchers in the conglomerate of the knowledge production industry when we are thinking about who drives the research agenda. It begs the question: What is the point of all this knowledge production if it is not impacting the researched? For whom is the research produced, and to what end?

Notes

1 For the thesis summarized in poetic form, see Ndlovu (2020).

References

Chilisa, Bagele, and Gabo Ntseane. 2010. "Resisting Dominant Discourses: Implications of Indigenous, African Feminist Theory and Methods for Gender and Education Research." *Gender and Education* 22 (6): 617–32. https://doi.org/10.1080/09540253.2010.519578.

Dorman, Sara, Jocelyn Alexander, Joann Mcgregor, and Terence Ranger. 2000. "Violence and Memory: One Hundred Years in the 'Dark Forests' of Matabeleland." *Social History of Africa* 72 (1): 173–74. https://doi.org/10.2307/3556810.

Dube, Luyanda, Maned Mhlongo, and Patrick Ngulube. 2014. "The Ethics of Anonymity and Confidentiality: Reading from the University of South Africa Policy on Research Ethics." *Indilinga: African Journal of Indigenous Knowledge Systems* 13: 201–14.

Erasmus, Zimitri. 2017. *Race Otherwise: Forging a New Humanism for South Africa.* Johannesburg: Wits University Press.

Gallagher, Charles A. 2000. "White Like Me? Methods, Meaning and Manipulation in the Field of White Studies." In *Racing Research, Researching Race: Methodological Dilemmas in Critical Race Studies*, edited by France Winddance Twine and Jonathan Warren, 67–92. New York: New York University Press.

Hadj-Abdou, Leila. 2019. "Immigrant Integration: The Governance of Ethno-Cultural Differences." *Comparative Migration Studies* 7 (1): 15. https://comparativemigrationstudies.springeropen.com/articles/10.1186/s40878-019-0124-8.

Hall, Stuart. 1990. "Cultural Identity and Diaspora." In *Identity: Community, Culture, Difference*, edited by Jonathan Rutherford. London: Lawrence and Wishart. muse.jhu.edu/book/34784.

Hendricks, Cheryl. 2018. "Decolonising Universities in South Africa: Rigged Spaces?" *International Journal of African Renaissance Studies - Multi-, Inter- and Transdisciplinarity* 13 (1): 16–38. https://doi.org/10.1080/18186874.2018.1474990.

Jacobsen, Karen, and Loren Landau. 2003. "The Dual Imperative in Refugee Research: Some Methodological and Ethical Considerations in Social Science Research on Forced Migration." *Disasters* 27: 185–206. https://doi.org/10.1111/1467-7717.00228.

Jayawardane, M. Neelika. 2019. "'The Capacity-Building-Workshop-in-Africa Hokum.'" *Journal of African Cultural Studies* 31: 276–80. https://doi.org/10.1080/13696815.2019.1630265.

Kalinga, Chisomo. 2019. "Caught between a Rock and a Hard Place: Navigating Global Research Partnerships in the Global South as an Indigenous Researcher." *Journal of African Cultural Studies* 31 (3): 270–72. https://doi.org/10.1080/13696815.2019.1630261.

Landau, Loren B. 2019a. "Capacity, Complicity, and Subversion: Revisiting Collaborative Refugee Research in an Era of Containment." In *Mobilizing Global Knowledge: Refugee Research in an Age of Displacement*, edited by Susan McGrath and Julie E.E. Young, 25–44. Calgary: University of Calgary Press. https://doi.org/10.2307/j.ctvpr7r1q.4.

———. 2019b. "A Chronotope of Containment Development: Europe's Migrant Crisis and Africa's Reterritorialization." *Antipode* 51: 169–86. https://doi.org/10.1111/anti.12420.

Leong, T. 2009. *Cities of Refuge Site Exploration-Musina.* Johannesburg: University of the Witwatersrand.

Lundström, Catrin. 2010. "White Ethnography: (Un)Comfortable Conveniences and Shared Privileges in Field-Work with Swedish Migrant Women." *NORA - Nordic Journal of Feminist and Gender Research* 18: 70–87. https://doi.org/10.1080/08038741003755467.

Lykes, M. Brinton. 1997. "Activist Participatory Research among the Maya of Guatemala: Constructing Meanings from Situated Knowledge." *Journal of Social Issues* 53 (4): 725–46. https://doi.org/10.1111/j.1540-4560.1997.tb02458.x.

Maldonado-Torres, Nelson. 2007. "On the Coloniality of Being." *Cultural Studies* 21: 240–70. https://doi.org/10.1080/09502380601162548.

Mbembe, Achille. 2016. "Decolonizing the University: New Directions." *Arts and Humanities in Higher Education* 15: 29–45. https://doi.org/10.1177/1474022215618513.

Mekgwe, Pinkie. 2006. "Theorizing African Feminism(s): The 'Colonial' Question." *QUEST: An African Journal of Philosophy / Revue Africaine de Philosophie*, Special Issue on African Feminisms, 20: 11–22. https://doi.org/10.1163/9789401205641_011.

Mignolo, D. Walter, and Madina Tlostanova. 2006. "Theorizing from the Borders: Shifting to Geo- and Body-Politics of Knowledge." *European Journal of Social Theory* 9 (2): 205–21. https://doi.org/10.1177/1368431006063333.

Moyo, Inocent. 2016. "Changing Migration Status and Shifting Vulnerabilities: A Research Note on Zimbabwean Migrants in South Africa." *Journal of Trafficking, Organized Crime and Security* 2: 108–12.

Nadar, Sarojini. 2014. "'Stories Are Data with Soul' – Lessons from Black Feminist Epistemology." *Agenda* 28: 18–28. https://doi.org/10.1080/10130950.2014.871838.

Ndlovu, Duduzile. 2010. *Migrant Communities' Coping with Socio-Political Violence: A Case Study of Zimbabwe Action Movement in Johannesburg, South Africa.* Johannesburg: University of the Witwatersrand, 1–84.

Ndlovu-Gatsheni, Sabelo. 2017. "Decolonizing Research Methodology Must Include Undoing Its Dirty History." *Journal of Public Administration* 52 (1): 186–88.

Ndlovu, Duduzile. 2020. "Imagining Zimbabwe as Home: Ethnicity, Violence, and Migration." *African Studies Review* 63 (3): 616–639. https://doi.org/10.1017/asr.2019.65.

Oliveira, Elsa, and Jo Vearey. 2015. "Images of Place: Visuals from Migrant Women Sex Workers in South Africa." *Medical Anthropology* 34. https://doi.org/10.1080/01459740.2015.1036263.

Palmary, Ingrid. 2009. "Migrations of Theory, Method and Practice: A Reflection on Themes in Migration Studies (Review Article)." *Psychology in Society*, 55–66.

Peberdy, Sally Ann. 1998. "Obscuring History: Contemporary Patterns of Regional Migration to South Africa." In *South Africa in Southern Africa: Reconfiguring the Region*, edited by David Simon, 187–205. Ohio: Ohio University Press.

Peberdy, Sally Ann, and Jonathan Crush. 1998. "Rooted in Racism: The Origins of the Aliens Control Act." In *Beyond Control: Immigration and Human Rights in a Democratic South Africa*, edited by Jonathan Crush, 18–36. Cape Town: Southern Africa Migration Project, IDASA (Institute for Democracy in South Africa).

Pécoud, Antoine. 2015. *Depoliticising Migration: Global Governance and International Migration Narratives*. Cham: Springer.

Pedri-Spade, Celeste. 2016. "'The Drum Is Your Document': Decolonizing Research through Anishinabe Song and Story." *International Review of Qualitative Research* 9: 385–406. https://doi.org/10.1525/irqr.2016.9.4.385.

Phoenix, Ann. 2008. "Analysing Narrative Contexts." In *Doing Narrative Research*, edited by Molly Andrews, Corrine Squire, and Maria Tamboukou, 72–87. London: Sage Publications.

Schinkel, Willem. 2018. "Against 'Immigrant Integration': For an End to Neocolonial Knowledge Production." *Comparative Migration Studies* 6: 1–17. https://doi.org/10.1186/s40878-018-0095-1.

Smith, Linda Tuhiwai. 1999. *Decolonizing Methodology: Research and Indigenous Peoples.* London: Zed Books.

Turner, Aaron. 2000. "Embodied Ethnography. Doing Culture." *Social Anthropology* 8: 51–60. https://doi.org/10.1111/j.1469-8676.2000.tb00207.x.

Turton, David. 1996. "Migrants and Refugees: A Mursi Case Study." In *In Search of Cool Ground: War, Flight, and Homecoming in Northeast Africa*, edited by Tim Allen, 96–110. Trenton, NJ: Africa World Press.

Vanyoro, Kudakwashe P. 2015. *Pragmatic Pathways: Critical Perspectives on Research Uptake in the Global South.* Migrating Out of Poverty. Brighton, UK: University of Sussex. http://www.migratingoutofpoverty.org/files/file.php?name=wp30-vanyoro-2015-pragmatic-pathways.pdf&site=354.

Vanyoro, Kudakwashe P., Leila Hadj-Abdou, and Helen Dempster. 2019. "Migration Studies: From Dehumanizing to Decolonising." *LSE Higher Education Blog*. https://blogs.lse.ac.uk/highereducation/2019/07/19/migration-studies-from-dehumanising-to-decolonising/.

Wee, Kellynn, Kudakwashe P. Vanyoro, and Zaheera Jinnah. 2018. "Repoliticizing International Migration Narratives? Critical Reflections on the Civil Society Days of the Global Forum on Migration and Development." *Globalizations* 15 (6): 795–808. https://doi.org/10.1080/14747731.2018.1446600.

Zaman, Shahaduz. 2008. "Native among the Natives: Physician Anthropologist Doing Hospital Ethnography at Home." *Journal of Contemporary Ethnography* 37: 135–54. https://doi.org/10.1177/0891241607312495.

3

Memorial Artwork of Deaths in Transit: Geographic and Temporal Imaginaries of Transnational Empathy

Danyel Ferrari

Introduction

On January 31, 2016, then Berlin-based Chinese dissident artist Ai Weiwei, who was running his studio on the Greek island of Lesvos, a noted point of arrival for asylum seekers crossing the Mediterranean from Turkey, posted a controversial image to his Instagram in which he is posed, limp, face down on the beach, recreating the infamous photograph of Alan Kurdi (Archey 2016; Dhillon 2016). The image of the tiny three-year-old Syrian-Kurdish boy's body washed up on the shore of the Turkish coast of Bodrum after his family attempted the same Mediterranean passage became, for a time, the international symbol for what was then troublingly and repeatedly termed "The European refugee crisis" (Slovic et al. 2017). Ai Weiwei had been using his popular Instagram feed to chronicle his experiences with asylum seekers and volunteers in Greece for some months prior to the recreation of the image. The image of Ai Weiwei face down in the sand was taken by the photographer as part of an interview for *India Today* on the artist's then-upcoming participation in the Indian Biennale, where the photo would later appear for sale (Jayaraman 2016; fig. 3.1). In the interview Ai Weiwei said he wanted to recreate the image with his own body to "feel" what could only be felt there, to "show solidarity," and "raise awareness" among his audience (Jayaraman 2016). Two weeks after the image was posted, he installed 14,000 life jackets he collected from Lesvos along with an inflatable boat emblazoned with "#SAFEPASSAGE" in hand-painted white letters on the columned facade of the Konzerthaus in Berlin as part of the 2016 Cinema

Fig. 3.1 Ai Weiwei reenacts the photograph of Alan Kurdi. *India Today photographer Rohit Chawla.*

Fig. 3.2 People walk past an art installation by Chinese artist Ai Weiwei that consists of life vests worn by refugees bound to the columns of the Konzerthaus at Gendarmenmarkt in Berlin. *Clemens Bilan/Getty Images.*

for Peace Gala event (fig. 3.2).[1] Images of the installation taken by passersby and covered by local and international media were "virally" reposted with articles and social media content videos calling the work a "powerful" statement of solidarity.[2] These and numerous other artworks appeared in the months that followed the summer of 2015. Many artworks, like Ai Weiwei's, were created by people outside of the refugee communities who claimed to be working to "raise awareness" of the issue among their audiences in contexts where migration could hardly be imagined to be an unknown subject. Like the term "European refugee crisis" (AJ+ 2016a; Holmes and Castañeda 2016), these artworks as awareness-raising projects have largely suggested that the increased number of people forced to migrate from primarily Syria and Libya, were all intent upon European destinations (a fallacy explored in detail in other chapters of this volume), and configured and imagined their audiences whose "awareness" they wish to raise as "European."[3] I argue that more than raising "awareness," they have sought to evoke an affective response of "feeling," particularly "empathy," from their imagined audiences and have done so by largely responding to and representing migration primarily through deaths in transit, utilizing various particular material, visual, and temporal constructions of memorials. This chapter explores the effects of such modes of affective appeal and the potential harms of the narrative geographies they rely upon and reinforce.

Critical to the framing of these projects has been the discourse of "empathy" as a political response to the crisis of forced migration. In interviews and videos created covering Ai Weiwei's various "activist/art endeavors," he repeatedly makes claims that the solution to the problem of the crisis of forced migration and deaths in transit is more "empathy" among European individuals (AJ+ 2016a). Ai Weiwei is far from alone in the use of this term. Empathy, as a form of political discourse and proposed solution to transnational issues of social justice, has, according to the scholar of media and affect, Carolyn Pedwell (2014b), "become a Euro-American political obsession." Often understood as an individual "capacity" for transnational intersubjective feeling, empathy has been touted as a panacea for social ills from inequity to explicit violence. The ability to "experience" another's feelings, particularly fear, pain, and other negative feelings, is then said to potentially transform the person engaging in empathy, arguably for the good of the person for whom they are experiencing empathy. But to what end? What kinds of subjects (and objects) are produced by public entreaties to empathy? On what terms does one become either a subject or object of empathy? What geographies of empathy are created?

This chapter argues that, contrary to challenging inequalities, immediate memorialization, its evocation of "transnational empathy" (Pedwell 2014a), and the narrative and visual tropes of neoliberal humanitarian "care" that undergirds such claims, work to elevate the image of Europe through expressions of individual feelings of "empathy" writ large as European values, while these same feelings have been evoked to allow the mortally unequal distribution of inclusion and mobility of neocolonial governmentality to continue, unabated and largely unquestioned. In order to build this argument, I begin by framing the recent expansion of both the political discourse of empathy, as articulated by Pedwell, and its unique paradoxical relationship to media and culture, configured as both a threat and possible solution to empathy as prosocial capacity.[4]

I then show how the rise of the political discourse of empathy has appeared in the arts as myriad new institutes and projects are dedicated to utilizing the arts to cultivate empathy. I consider how such calls for art to function as a tool of "empathy" capacity-building or "empathy boot-camp" (Musiker 2015) often with little critical analysis of the potential effects of empathy beyond a general "prosociality" can be critically understood to draw upon a history of related antecedent notions of "sentiment" essential to hegemonic humanitarian visualities. In the absence of scholarship focused on the social effects of art as a tool of empathy building in awareness-raising artworks, I look to the considerable work of anticolonial and critical humanitarian studies scholars (Malkki 1996; De Genova 2017; Ticktin 2011, 2016; Perera 2013), and their critiques of how dominant humanitarian narratives and the affects and visualities they produce have not only reified the distinctions of the subject of humanitarian sentiment—those who *feel*, and its objects, and those *for whom* it is felt—but have done so along hegemonic, neocolonial geographic imaginaries (Ticktin 2016). These dominant narratives suggest that not only is global migration problematically and inaccurately imagined to predominantly originate in the Global South and East and "flow" toward destinations in the North and West, but that the complementary trajectories of transnational humanitarian sentiments or empathy that influence the artworks discussed here are imagined coming from the Global North, directed to the Global South, or, "from the West to the rest."[5]

Finally, thinking through previously explored race- and gender-informed critical humanitarian scholarship on humanitarian visualities, I analyze these awareness-raising artworks' utilization of specific material and visual strategies like "reenactment," as immediate memorializations, interrogating how they work as projects of transnational empathy to construct their subjects/

objects and enact geographic and temporal imaginaries of apprehension and knowledge, proximity, and distance.

The Rise of the Political Discourse of Empathy

Empathy, used colloquially to mean the ability to "feel another's feelings" or, idiomatically, to "walk in another's shoes," has, according to Carolyn Pedwell's (2012, 2014a, 2014b, 2016) extensive work on the subject, become something of an "obsession" in American and European political discourse. In her book *Affective Relations: The Transnational Politics of Empathy* (2014a) and related work, Pedwell chronicles the increase in the use of empathy in political communication, particularly since the early 2000s, and considers its potentials and possibilities for genuine solidarity and its capacity to reinstate hegemonic subjectivities. She notes that in political discourse, its development and practice as capacity is elevated beyond the individual to a social "good" suggested as something of a panacea to numerous and varied social ills (Pedwell 2016, 28).

For all its sudden ubiquity and value, the term "empathy" is relatively new to English. The word has roots in the arts and modernist aesthetic appreciation and came into being only in 1908 when it was coined as a translation for the German "Einfühlung" (in-feeling).[6] "Einfühlung" is used to describe the "aesthetic activity of transferring one's own feeling into the forms and shapes of objects" (Lanzoni 2018, 16). Einfühlung/empathy was not conceived to describe its current use as the ability to "feel another person's emotion," but, rather pointedly, its opposite. To enact empathy, in the early 1900s, meant to "enliven an object, or to project one's own imagined feelings onto the world" (Lanzoni 2018, 16); in other words, to expand the subjectivity of the viewer. In the mid-twentieth century, the term began to shift toward the psychological and social, due in part to the postwar popularization of psychology. "Empathy" was perceived to be more scientific sympathy (Lanzoni 2018, 20). Empathy was believed to offer new possibilities for connection, identification, and understanding that might improve social relations of all kinds, contributing to its early uses across psychology, advertising, education, and politics. In the 1960s this "aspirational value" began to be imagined to be able to help build tolerance and even to "eradicate prejudice" (Lanzoni 2018, 24).

The merging of the individual psychological and the social justice understandings of empathy continues to be a foundational assumption of the prosocial value of empathy in contemporary popular appeals. Pedwell identifies two groups with different modes of understanding empathy and its political effects: first, there are the theorists who understand empathy to be primarily a "capacity, skill, or tool" (2012, 284), and secondly, there are those that consider empathy "as a social relation or product of circulation" (Ahmed 2004; Berlant 2004). The latter often emerge from affect theory to correct an ahistorical and depoliticized view of emotion as "individual" and are interested in the potential political effects of empathy (and its antecedents) beyond the unexamined "social good" imagined in the capacity framework. For instance, Sara Ahmed (2004, 29) notes that "empathy sustains that very difference that it may seek to overcome" when subjects assume that they can feel what another feels in ways that fail to take account of differences in history, power, and experience. The majority of popular culture and political discursive uses and understandings of empathy, particularly those evoked in the artworks here, consider it from the previous understanding, as a personal capacity, to be either lost or gained depending on "habits" of (dis)engagement, and largely critiqued for its actual effects on social relations.

Recent calls to individuals to build their capacity for empathy in order to be more socially engaged emerge amid a societal diagnosis of a decline in empathy. Numerous books and studies have emerged stating that empathy is precariously under threat. A commonly stated cause for its imperilment is media consumption, and in particular, social media usage.[7] Such a threat not only speaks to anxieties over interpersonal relations but the concept of Western liberal democracy, more broadly. The obsession with empathy in Euro-American political discourse, noted by Pedwell, can be understood as a response to a crisis or moral panic over a loss of a social capacity identified as critical to understandings of human rights as rooted in Western moral sentiments (Fassin 2011). Paradoxically, in response to such concerns, numerous attempts to engender greater or better empathy have appeared, particularly through media.

The image of Kurdi that Ai Weiwei reproduced was largely shared online as an "awareness-raising" tool understood to engender empathy among transnational audiences and has since been studied as an image related to the capacity to engender, hold, and focus empathy toward social change (Slovic et al. 2017). Similarly, numerous humanitarian aid media projects that emerged in response to the crisis of forced migration have largely been built around the concept of expanding the empathy of the viewer for the refugee by putting the viewer in their "shoes." One popular video by Save the Children called "Second a Day," for example, tried to "bring the crisis home" by putting a white, European child in the context of forced migration, telling the story of her life from one birthday to the next; during that year her life is transformed from a middle-class life of comfort to loss of home and sickness intended to suggest the experience of refugee children from Syria (Save the Children 2016).[8]

The Art of Transnational Empathy and the Visualities of Suffering in the Context of Securitization

In the paradox of empathy's relationship to media, art has come to be considered as a particular form of media that can counter the harmful effects of social media and other forms of media that threaten the capacity for empathy. Numerous articles have appeared in recent years, not only citing the previously stated deficit of empathy but also touting the arts as uniquely suited to correct the issue. A sample of the titles not only belies the idea that this is a special purview of the arts but also echoes Pedwell's assertion that it is now understood primarily in public discourse as a skill essential in Euro-American neoliberalism: "Can Art Help People Develop Empathy?" (Caldwell 2018); "Fighting the Empathy Deficit: How the Arts Can Make Us More Compassionate" (Musiker 2015); and "Using the Arts to Build Empathy, the *Ultimate* 21st Century Skill" (Moore 2013, emphasis mine). The School of the Art Institute of Chicago published an article which phrases it thus:

> Artists can increase empathy in others through their work, eliciting that feeling from people who may be numb from all the terrible things going on in the world, making the viewer more sensitive and vulnerable (Logan n.d.)

Numerous talks, exhibitions, and special projects[9] have appeared in art institutions in the United States and Europe to address empathy and explore art as a means of increasing capacities for it. For example, The Stanford Art Center hosted an exhibition and related lecture and course titled *Empathy* that argued for the value of visual art in the production of empathy and "universality." In addition to talks and articles touting this unique capacity, projects and programs have emerged to pursue the production or "training" of empathy as a central mission of emerging museums and research projects. In 2016, the Minneapolis Institute of Art announced its development of the Center for Empathy and the Visual Arts (CEVA n.d.), funded by a partnership of transnational corporate cultural foundations, including the Carnegie, Ford, and Walton Foundations. CEVA's stated mission tasks itself with working to make its audiences better at empathy through experiences with art, implying that this process will solve not only relationships between individuals but between "groups" of people (CEVA n.d.).[10]

The arts have not only been suggested as a localized or individual tool of prosocial relations but have been specifically posited as a tool of international human rights. In 2017, the European Union Agency for Fundamental Rights (FRA) published its first "report of high-level expert meeting," *Exploring the Connections between Arts and Human Rights*. The report specifically noted that a critical point of overlap between the arts and human rights models surrounds "universality" and its ability to evoke through empathy and "feelingful thinking":

> Both are concerned with questions of what is (and what is not) humanity, identity, dignity, of communicating empathy, of the transformation of lives, of visions for the future and of the mission of mankind, of the full development of the person. Both are universally applicable (2017, 6).

While the political discourse of empathy and the arts' potential for its development is relatively new, the concern for social feeling in response to representations of suffering and the idea that media is, paradoxically, both the cause and possible solution to a lack of such feeling is far from new. The importance of the visual representation of atrocities in order to draw the sentiments and interests of those in power is deeply ingrained in our understanding of visuality and humanitarianism. These artworks as awareness-raising appeals via large-scale memorial public artworks and related far-reaching social media content evince a novel position for mediatized art as a new vehicle of humanitarianism. While there is scant critical analysis of empathy discourses in the various calls to consider art as a means of engendering it, empathy, as it is evoked in these artworks, can be considered as part of a larger history of humanitarian visualities.

Humanitarian appeals have long been tied in a complex relationship to the visual and to a representation of death and suffering. Susie Linfield (2010) has written in *Cruel Radiance* that human rights images are particularly problematic in so far as rights cannot be visualized except through their absence:

> Philosophies that undergird ideas about human rights are . . . built around absence. And photographs, I would argue, are the perfect medium to mirror the lacunae at the heart of human rights ideals. It is awfully hard to photograph a human right . . . in fact rights don't look like anything at all . . . What a photographer can do, and do

particularly well, is to show how those without such rights look, and what the absence of such rights does to a person (Linfield 2010, 37).

Following such logic, humanitarian claims are then based upon the recognition of an absence, and as such, many of the images produced around them are based on the demonstration of evidence of exactly that absence, leading to troubling consequences. Humanitarianism's particular use of sentiment has been widely and roundly considered by anticolonial, critical race, and migration scholars alike for the violent biopolitical effects of its representations of suffering subjects and its valorization of interventionist imperialism. While public art that utilizes memorial within seemingly activist appeals has often been met with less skepticism, there is no shortage of scholarship on the negative effects of affective appeals made on the terms of representations of suffering and death.

Critical race-informed humanitarianism studies scholars, in particular, have addressed broader themes of appeals to visuality applied to the subject of humanitarianism. Most notably, Malkki (1996), Ticktin (2016), Perera (2009, 2013), and Rajaram (2002) all suggest that the ways in which suffering must be made visible in order to elicit sentiment in humanitarian appeals are problematic. Malkki and Rajaram identify the ways in which visibility models built around vulnerability produce the humanitarian subject as a "speechless object." Referencing Giorgio Agamben's concept of "bare life" (1995), Malkki (1996) writes that the act of depicting suffering not only depends upon the lacunae of human rights, as identified by Linfield, it also *produces* the objects of those rights discourses as "speechless emissaries" and figures of pure suffering, in order to make them the "pure" victim necessary to make the claims of need required for intercession visible. In so doing, the structure of humanitarianism and the visualities it gives rise to actually work to make the figure of human rights unrelatable as anything more than the object of sentiment. Following Malkki, Prem Kumar Rajaram (2002), in his work on humanitarian postcards sold by the United States Committee for Refugees (USCR) and Oxfam, has suggested that these images, in particular, commodify bodies and experiences of humanitarian subjects as a means of raising funding.

Humanitarian visualities, as applied to asylum seekers, function somewhat differently from other forms of humanitarian visuality in that the "other" is not necessarily physically distant, and the appeal is not to send aid *elsewhere* but, often, to offer inclusion, however limited. Migration scholars Perera (2009) and Ticktin (2016) apply critical analysis to humanitarian visualities' required reliance on suffering as a measure of inclusion in the international refugee regime (as noted in great detail in the work of Didier Fassin), which works to demand unattainable demonstrations of innocence and vulnerability. They build on Malkki by adding that requiring refugees to be "silent objects" of suffering, in turn, discursively constructs any form of agency as an "excess" by which "deservingness" can be countered and denied. As Ticktin (2016, 261) notes,

> Perversely, innocence is clearest in death; Alan exemplified the pure innocence of refugees only after he was washed up on Turkish shores. He was pure in his passivity, his lack—innocence is about lack, after all.

Ticktin points out that the totality of the humanitarian regime has often been presented as the only option for those opposed to right-wing arguments of total exclusion, and indeed, humanitarian visualities are configured to contend with the discourse of the nation's securitization. Ana Ivasiuc (2018, 235) terms the visuality of securitization "the complex of securitarian visuality" in her exploration of vernacular visualities on social media by anti-Roma groups in Rome. She articulates her theorization in conversation with multiple concepts of visuality, particularly responding to visual culture scholar Nicholas Mirzoeff's "complex of visuality" (2011). Drawing on Foucault, the "complex of visuality" is defined as an "imbrication of mentality and organization produc[ing] a visualized deployment of bodies and training of minds, organized so as to sustain both physical segregation between rulers and ruled, *and* mental compliance with those arrangements" (Mirzoeff 2011, 7, emphasis mine). Ivasiuc credits Mirzoeff's genealogy of visuality as emerging from power dynamics of the mid-seventeenth century and working to construct itself as "natural" while constituting power dynamics. She importantly contends with his understanding of how security functions within this complex. Whereas Mirzoeff describes "the body-politic" as "so enmired in 'security' as to have lost a sense of purpose" (2011, 283), Ivasiuc (2018, 253) suggests that Western societies, in fact, fabricate a sense of purpose—and of community—precisely around security, which has become a central principle of social ordering.

Applying Ivasiuc's view to humanitarian visualities, these visualities of suffering, of a competitive vulnerability (of which, arguably, mortality is the highest measure, as noted by Ticktin earlier), can be seen as a rebuttal to the representation of refugees in right-wing anti-immigration messaging configured as "invaders" or "threats." Following these assumptions, the method to gain support for entry and inclusion, as seen in empathy and "awareness-raising" memorial public art, is conditioned by, even if in opposition to, the security complex. As a result, the terms of these appeals are often too easily reframed to further support securitization tactics. As Ticktin (2016, 259) notes, ultimately, "innocence works as part of a binary: guilt is its necessary other, and the pendulum can swing quickly between the two." Those who seek not to open borders nor to guarantee safe passage but instead to justify increased policing of existing borders and the waters that surround them can and do similarly cite these deaths as cause for their goals. Migration scholar Nicholas De Genova (2017) writes:

> [T]he invocation of tragedy more or less came to be cynically conscripted to supply the pretext for reinforcing and aggravating precisely the material and practical conditions of possibility for the escalation of migrant deaths—namely, the fortification of border policing that inevitably serve [*sic*] to channel illegalized human mobility into ever more perilous pathways and modes of passage. (7)

Frontex, the EU's external border agency, has repeatedly stated that increased "border control" tactics, and even explicit violent attacks on migrant boats, are in the interest of discouraging more attempts at passage in order to "protect" vulnerable refugees from "smugglers" and "traffickers."[11]

The logics of humanitarian visualities reinforce fictive dominant narratives of trajectories that center Northern Hemispheric and Western destinations and do not call into question the legitimacy of borders or even of securitization strategies but instead depend upon a vulnerability sufficient enough to argue at best for an *exception* that ultimately supports the rule, as argued by Hannah Arendt (1979, 299). By demanding the asylum seeker must present as vulnerable in order to be visible as lacunae of rights so that he or she may prove to be "deserving" of the intercession of another (European) state, the individual is not only stripped of any humanity or agency to become "bare life," but, in addition, the structures appealed to for intercession are simultaneously elevated and constructed as "saviors."

The figure of the "other" seeking entry reconfigured primarily through limiting representations of suffering confirms the construction of the organization, institution, nation, or even international body (in this case, the EU) that is being appealed to on his/her behalf. The destination state of migrants rhetorically reaffirms its value via the figure of the refugee. In a neoliberal society wherein capital and goods flow across borders and corporations are transnational, the reaffirmation of the nation-state in these contexts takes on particularly affective resonance. Wendy Brown (2010) has detailed how the "wall"—the ultimate symbol of the border—has become increasingly significant as the economic sovereignty of nations has waned in the neoliberal era. The "spectacle" of forced migration, and even of the deaths and violence suffered therein, in particular in photographic form, detailed by De Genova (2017), works to maintain not only the validity of the nation but its desirability as such.

De Genova, drawing upon Guy Debord's work, suggests that the "spectacle" of migration (always configured as South to North/East to West) works to reify Europe as both an idea and an ideal destination. Its most extreme spectacle, that of death in transit, is discursively turned into a "sacrifice" to the democratic specter of the EU or "Europe."

Connecting the thinking of Malkki, Ticktin, and De Genova, the refugee who dies in transit becomes the symbol for all that is valuable and desirable in Western "democracy." Among the living, those who succeed in surviving transit and manage to gain entry must demonstrate *sufficient* vulnerability and show themselves to be in such need that they are "grateful" and open to being "reeducated" to "European ways" in order to no longer carry the specter of threat with them. Their "acceptability" as a discursive figure of "refugee" (rather than threat) relies on how they reaffirm the value of the host nation, either through sacrifice or perpetual gratitude and continued vulnerability.

Immediate Memorialization and the Temporal Geographies of Transnational Empathy

Ai Weiwei's embodied recreation of the image of Kurdi's death is fundamentally an act of reenactment as memorial. Reenactment has emerged as an artistic and historical/memorial strategy in recent decades as a means of both "embodied" or "affective" education. The affective desire for a haptic connection to the past compels reenactment as a means to "narrow the gap between past and present so that we might touch it" (McCalman and Pickering 2010, 8). Through such an understanding, reenactment emerges not merely as an attempt to describe history but as an affective gesture toward an "embodied" apprehension of it, an enclosure around it. Reenactment offers a means of approaching the event boldly, in other words, affectively, a gestural means promising the ability to "step into another's shoes." In her introduction to the exhibition catalogue *History Will Repeat Itself,* art critic and curator Inke Arns (2007) defines the fundamental aspect of all reenactment-based artworks as their paradoxical relationship to the distance between the original image and the viewer. She states that the "erasure and simultaneous creation of distance are two key mechanisms in the contemporary practice of artistic re-enactments, which often coexist in one and the same artwork" (Arns 2007, 59). She writes:

> Re-enactments eliminate the distance, constructed as safe, between the historical event and represented by the media and the present, between performers and audience. The re-enactment transforms representation into embodiment, distanced indirect involvement into—sometimes unpleasant—direct involvement, and through this turns the passive reader or observer into an active witness or participant (2007, 59).

As an action, reenactment works to close the distance between the viewer and the image through embodiment. Ai Weiwei purportedly intends to do just this when he lies down, to close the distance between himself as a viewer of the image and also between his audience and the image. This is arguably an affective gesture intended to close both an affective and a physical distance through the use of his body. In the interview for *India Today* that produced the image, Ai Weiwei describes his motivations as follows:

> I was standing there and I could feel my body shaking with the wind—you feel death in the wind. You are taken by some kind of emotions that you can only have when you are there. So for me to be in the same position [as Kurdi], is to suggest our condition can be so far from human concerns in today's politics (Jayaraman 2016).

But, as Arns suggests, the closing of distance is only one effect of the reenactment; as a strategy of history, reenactment does not only *close* a temporal distance, it creates another, as well. Arns argues this is always the second effect of reenactment, that paradoxically, reenactments, as artistic gestures, produce distance as well as close it. Art reenactments do not only haptically recreate; they, by necessity, create some kind of *image* from the embodied actions and, with it, an additional distance of mediation, or in this case, "re-mediation." In the context of artworks addressing historic moments, this distance is productive of new imaginings of history and awareness of the complex systems of mediation in which history is already embedded and through which it is produced. In the context of the immediate re-creation of an image that was very much of the moment, the re-mediation does something decidedly different. At the time of the images' release, art critic and curator Karen Archey criticized Ai Weiwei's photograph as a callous replacement of one body deemed less valuable with another, more culturally valuable body. She writes:

> [T]he artist fails to consider that by posing as Alan Kurdi, he effectively suggests that the image of a dead three-year-old isn't sufficiently shocking to us, that we need instead an image of a middle-aged, wealthy and powerful man to confer the real tragedy of the refugees' plight (Archey 2016).

By replacing the body of the boy with his own, not only is Ai Weiwei feeling something through an embodied action, but he is producing an image, which, as Archey posits, suggests that the distance between the viewer and the original image would be closed by our more immediate identification with the body of the wealthy artist over that of the drowned toddler.

Another distance is created in the moment of Ai Weiwei's reenactment. If the distance closed in *most* reenactment art is temporal, making the past present by nature of recreating an historic event, what is the effect when the event being recreated is current? What happens instead when the memorial artworks are produced a mere six months after the death of the child when the image had hardly disappeared from the media by the time he reenacted it? This was an *immediate* reenactment, an immediate memorialization, even as the deaths continued. Arns notes (2007, 45) that the purpose of a reenactment is both to access the history in the present and to facilitate critical thought about processes and mechanisms of historicity. But in the gesture and the resulting image, Ai Weiwei does not so much *criticize* the structures of historicity as enact them. As a subject of reenactment, the original image becomes an "historic" event; the image of the original event, the death of Kurdi, is effectively pushed into the past while it is remade in the present.

This effect—the enclosure of events that signifies ongoing conditions in the past—is not unique to the reenactment memorial alone but appears in various strategies of memorialization in these larger-scale artworks. In her 2006 "Rushing to Memorialize," Janet Donohoe sets up two distinct models of "immediate memorialization" with distinct political effects. The ad hoc and immediate, typically enacted by the personally affected grieving community, is more of a public mourning than a memorial proper, in the sense that it is predicated on the life of an individual, not the total representation of an *event* in which their death is circumscribed. By contrast, the official immediate memorial is typically enacted by the state, in which the act of memorialization encloses deaths within an event and thereby codifies a particular narrative around that event. What is at play when those "immediately memorialized" on an "official" scale are "outsiders" for whose deaths the state is arguably culpable? Memorial awareness-raising projects seem to collapse distinctions between mourning and the memorial. The political significance of who has a right to be mourned and who does not has been well attended to by scholars such as Judith Butler. This premise, that the nation (or we might also extend this to extranational communities) only mourns those that they value and excludes those they do not from rites of mourning and, by extension, memorialization, precludes critical analysis of how memorializations can, themselves, as a form of cultural technology, enact violence through another model of a political, temporal imaginary, different from that defined by the far-right.

Large, public, institutional memorializations are decidedly more than private mourning scaled to the crowd. There are particular effects to having state, institutional, or corporate stakeholders enact memorializations of those decidedly not included in the nation, and there are even more complex implications for their doing so even as the events and conditions that contribute to those deaths continue. Anthropologist and theorist of late liberal neocolonial governmentalities Elizabeth Povinelli (2011) explains in her *Economies of Abandonment* that the systemic inequalities of what she has termed "late liberalism"[12] are sustained through various imaginaries that make unequal global distributions of "life and death," "hope and harm," appear "sensible" (3). Povinelli begins her consideration of economies of abandonment with the parable, "The Ones Who Walk Away from Omelas" borrowed from writer Ursula Le Guin (1976). In the story, Omelas is a fictional land without pain or illness, where everyone is happy and fulfilled, but where the cost of this perfect happiness is the endless suffering of one child, confined to a broom closet of which all must be made aware. This story is, for Povinelli, a touchstone of late liberalism but with two critical distinctions: unlike Omelas, she explains, late liberalism as a neocolonial project inflicts suffering on the *many* for the happiness of the *few*, and it is a unique temporal imaginary that allows for this. According to Povinelli, one of the key imaginaries critical to how late liberalism aggregates social worlds is through tense. The temporal imaginary of the future anterior or "future perfect" promises a *future* happiness of the many. Through discursive temporal imaginaries, the present moment is imagined as if from a "future anterior" view—a "looking forward to look back" from an imagined future wherein all are equal, and all suffering up to that point is transformed into a necessary, justified sacrifice:

> The ethical nature of present action is . . . interpreted from the point of view of a reflexive future horizon and its cognate discourses, such as that of sacrificial love . . . Suffering disappears when seen from the perspective of what it will have been—or been for (2011, 5).

Povinelli does not immediately address memorials per se. She is, however, interested in the ways in which "sacrificial love" appears and is codified within "economies of abandonment"[13] to justify various acts of violence in the fractured present. I am contesting that memorials of ongoing deaths caused by the unequal distribution of "hope and harm, and of endurance and exhaustion in late liberalism" (Povinelli 2011, 3) are precisely one of the cultural technologies of "tense" through which imaginaries of tense are materialized and made sensible.

Shortly after Ai Weiwei's #SAFEPASSAGE went viral, a less well-known artist, British artist Jason deCaires Taylor, installed his *Raft of Lampedusa* (fig. 3.3) off the coast of the Canary Islands. The artwork's title is a reference both to the Italian island where an estimated 360 Libyan migrants died in a shipwreck in 2013 and to Théodore Géricault's famous painting *Raft of the Medusa* (fig. 3.4), which depicted the aftermath of the historic wreck of the French naval frigate *Méduse* and the deaths, cannibalism, and other horrors that followed. Taylor's sculpture entailed a massive multifigure sculptural installation of a sunken inflatable boat, not unlike the one included in Ai Weiwei's #SAFEPASSAGE, but here the raft is heavy with bodies, some who appear to be deceased, and surrounded by many more resolutely walking along the ocean floor toward an unreachable destination, frozen in time.

Curiously, Taylor has resisted the reference to the piece as a memorial and instead has focused on the environmental aspects of his underwater concrete sculptures made from pH-neutral stone designed to serve as scaffolding for coral colonies. In a video produced for AJ+ about the piece, Taylor had this to say:

> The work is not intended as a tribute or memorial to lives lost but as a stark reminder of the collective responsibility of our now global community (AJ+ 2016b).

Fig. 3.3 The Raft of Lampedusa. *Photograph: Jason deCaires Taylor.*

Fig. 3.4 Théodore Géricault, *Le Radeau de la Méduse* (Raft of the Medusa), 1818–1919. Louvre Collection.

While Jason deCaires Taylor may well have not intended to produce a memorial, it would be difficult to argue that the work does not function as one. Not only is the material choice of figurative sculptures to scale (life-casts from models, themselves often migrants) reminiscent of traditional monuments but also the artist has previously made work that has been understood as memorials to historic violence. A past project, *Vicissitudes*, installed on the bottom of the Caribbean Sea in 2007 depicting figures holding hands in a circle facing outward, has been understood, if not initially framed, as a memorialization of those slaves who were thrown overboard during the Middle Passage. The artist has repeatedly asserted otherwise, focusing instead on the environmental aspects of the work as an intended future scaffolding for sea life. Despite the artist's statements, both of these works are repeatedly presented on the artist's own website in videos set to melancholic piano music and are titled "To Recollect Loss." Taylor, a self-proclaimed scuba diver and environmentalist, has instead focused on the environmental futurity of his several underwater installations as part of a larger project intended to create artificial scaffolding to support coral regrowth and thus support other sea life.

While his early work lacked the echo of memorialization in favor of depictions of stereotypical waste and excess of Western consumer culture, his later work took a decidedly memorial turn. Video updates on *Vicissitudes* show its figures overgrown by the nonhuman ecosystem they now support; the concrete of which they are constructed is often cracked and broken as the coral takes root. *Vicissitudes*, and by extension of temporal imagining, *The Raft of Lampedusa*, configure their subjects, the enslaved and refugees, respectively, as bodies disappeared through death at sea, reproduced in their absence as concrete stand-ins cast from other bodies. They become tombs to "Unknown" and unnamed dead, remade as nature. However, distinct from Benedict Anderson's tomb to the Unknown Soldier, they do not represent the ghosts of the nation lost in valor. Instead, however much the work intends to speak to responsibility, materially, the bodies are reconfigured as part of the earth—returned to the sea, "naturalized," and installed in their watery home. In reference to his sculptures, Taylor has said, "Once we submerge them, they don't belong to us anymore" (Taylor 2015). Susan Smillie, writing for *The Guardian* in conversation with art critic Jonathan Jones, writes this:

> "It's strange," he concludes. "There's a sort of dreamlike, redemptive poetry to it." If you consider that the sea is already a museum littered with artefacts and remnants—wrecks of Carthaginian ships, ancient Greek statues depicting heroes, warriors and gods—it begs

> the question: what will future generations make of our modern world as imagined by Taylor? . . . This benighted raft of the hungry and hopeless. "Maybe he's dreaming of a time where humans have been left behind, in a nature that's survived us," offers Jones. "We might be the forgotten ones" (Smillie 2016).

Jones's troubling identification of the *Raft of Lampedusa* as "redemptive" relies on a complete dehistoricization and depoliticization of the Libyan refugees depicted. Not only are they configured, in Malkki's and Rajaram's terms, as "hungry and hopeless," but in his projection of Taylor's imagination, they are reconfigured as already ahistoric, primordial representations of "humanness," in a projected future after humanity. In both Taylor's piece and Jones's statements, there is a barely concealed desire for abandonment, for the forgetting that accompanies narratives of redemption in memorials. Through Jones's narrative imaginings, he projects himself and the reader, in Povinelli's term, into a "future anterior" where we experience the sculptures (and, by extension, the deaths they represent) as looking backward from a future where they are already configured as past, their deaths no longer recent, suddenly farther away. He creates a "redemptive" future where he and his reader have moved past culpability.

Conclusion

Looking at awareness-raising memorials through their affective visual discursive strategies within the context of humanitarian visualities and particular temporalities shows us a complex and problematic effect of transnational empathy. While the experience of empathy can expand the subjectivity of the individual feeling it and the ideological or moral capital of the nation (or transnational entity) to which they belong, it often does so by reifying fictive geographies and subject positions that enable the continuation of the very neoliberal and neocolonial models of inequity its advocates purport it has the potential to address. As we have seen, empathetic sentiment around death in transit, articulated through awareness-raising artworks, while seemingly a "positive" response against nationalist anti-immigration rhetoric, is not unproblematic. Empathy's primary value to the subjects of feeling and not those for whom it is felt is evidenced by its "flexibility." Empathy is not immune to recapture and use by the border regime to argue for the protectionism that both increased border securitization and the brokering of the troubling EU/Turkey deal while projecting a Western Northern European (self)image of humanitarianism. These gestures of transnational empathy allow for the Western liberal state and its subjects to profess ideologies and sentiments without interrupting their "sovereign" mechanisms of neoliberal and neocolonial governmentality, justified as "securitization" and even protection for refugees. In part, empathy can work this way because of its reliance on visualities that produce the object of humanitarianism as always a suffering other, geographically originating "outside" of Western/Northern Hemispheric centers of liberal democracy. The geographic imaginaries of transnational empathy thus configure suffering as flowing *at* Europe from the outside (with no consideration of the conditions produced by Western interventions) and humanitarian aid, and the sentiments that undergird it, like the discourse of rights, flowing *from* it. Finally, as cultural technologies that produce imaginaries of tense as well as geography, these awareness-raising artworks construct a means to both draw close the object of humanitarian aid— through a form of "imperial knowledge," "reticulated knowledge" reticulated as "care for 'the other'" that pulls the "other" closer— and temporally construct their suffering as sacrifice to a "perfect" future that justifies, neutralizes, and allows various state violence and abandonments to continue, unimpeded.

Most transnational feminist and queer activist scholars, in one way or another, posit that empathy can, under the right conditions, build connections with potentials for political solidarity. Pedwell (2014b), too, holds a potential for genuine solidarity of empathy, but she notes in order for this to be possible, it must involve a "radically unsettling" affective experience emerging from a "complex and ongoing set of translational processes involving conflict, negotiation, and imagination." I remain slightly more skeptical of the ability of empathy, as it has come to exist within transnational flows of power and subjectivity production, to do this work of unsettling. Artworks organized around empathy calls in response to depictions of suffering and death of "others" fail not only to destabilize the self, but in so framing these others, they, like humanitarian visualities more generally, foreclose the possibility of emotions and identities beyond suffering. I am curious instead about the political possibilities of all those performances of self, emotion, identity, etc., which are perceived to be in "excess" to the vulnerabilities demanded of humanitarianism and art built on its same affective structures. What political possibilities might exist if we look beyond the models of moral sentiment and its reliance on the visibilization of vulnerability and its reification of power? How might this do more than destabilize an individual sense of subjectivity in viewers to resist the imaginaries that allow late liberal and neocolonial abandonments to appear just?

A critical call from refugee communities has been "nothing about us, without us." Self-representations of refugee community experiences work to resist modes of totalization of the community and of the fictive narrative tropes too often recited in calls to empathy based on suffering. A collective of activists, LGBTQIA+ Refugees Welcome (LGBTQIA+ Refugees 2017), has recently offered a strong guiding example. The group had been invited by Spanish artist Roger Bernat to participate in a project he was constructing as part of Documenta 14. The quinquennial, typically hosted in Kassel, Germany, included a portion of the 2017 Documenta 14 in Lesvos. As part of an enactment of "mass theatre" he had created with dramatist Roberto Fratini, *The Place of the Thing*, Bernat invited different groups and collectives in the city to carry a fiberglass prop of an ancient Greek monolith known as the "oath stone" through the city (Bernat 2017).

Once the project was "activated" by the hands of multiple groups, it would then migrate to its final destination in Kassel by plane. On May 21, 2017, the group of activists responded by "stealing" the central art object/prop of the project and releasing a press release, half manifesto, half ransom note, and an accompanying video, simultaneously scathing and joy-filled, in which the masked members collectively read their statement as a form of poetry, danced with the rock, humorously titled their intervention *Between a Rock and a Hard Place*, and rechristened the quinquennial "Rocumenta 14." In their video statement, they condemned Bernat's piece as exoticizing and stated, referring to the stipend they were offered, that they could not be bought, and that they were keeping the rock (LGBTQIA+ Refugees 2017). They further added, in reference to the funerary aspects of the work that invited participants to treat this inanimate object almost as a memorial to all of humanity: "You have asked us to perform a fake funeral for your stone, we've had more than our fair share of funerals" (LGBTQIA+ Refugees 2017, 1:43). In their mix of rage and joy, recrimination and humor, these artist/activists enact everything that is in "excess" of the required visibility of vulnerability, of suffering, in excess of the memorializing impulse. It is a loud and profound resistance to the desire to contain refugee deaths and construct a temporally bounded "event" from the continued conditions of forced migration and state abandonment.

In a curious turn, Bernat, who initially released a furious statement of indictment of the group and defense of his intentions, later reframed his response, saying that not only did he not mind but thanked the group for finally activating the artwork and letting them know he had an additional copy of the stone to carry on with the project. In that gesture, he attempted to reincorporate their action back into the regime of the sensible that would have not only "absolved" them but reasserted his benevolence and the "genius" of the piece, as if it had been conceived to foresee and encompass just such an action. It does something else important to this exploration; it enacts (or attempts) a form of "empathy" and demonstrates not only the previously discussed appropriative nature of it but another aspect we have yet to consider. How does empathy of this kind work to erase or surpass conflict and antagonism from those we might claim to empathize with by enclosing it in this way? It attempts to reaffirm not only the power dynamics at play that position the empathizer as a feeling subject and the empathized as an object but also the moral "correctness" of the subject.

The antagonisms offered by LGBTQIA+ Refugees in Lesvos, the resistance of the subject/object constructions and their terms, and the literal interruption of the reenactment of the circuits of transit being reified in the piece, pose an important challenge to the logics of awareness-raising memorials and their claims of affective resonance. The experience of watching their video statement/action/performance is not unemotional. It is, however, not an experience of consuming or an appropriation of "feeling another's pain," or even identifying a point of shared vulnerability. It is a feeling of seeing the sensible shift at the hands of those being represented. It is a resounding insistence of irreducible presence against the enclosures of imaginaries of tense and geography enacted by humanitarian awareness-raising memorials and an artwork reproducing them.

Notes

1 A related "action" at the Gala for Peace plated fundraiser dinner involved Ai Weiwei asking attending celebrities to don gold mylar thermal blankets like those sometimes given to migrants arriving on the shores of Lesvos for a mass "selfie moment." See Loughrey (2016).

2 See the Global Citizen tweet, 2016.

3 As with humanitarian aid organizations' awareness-raising videos, "European" has appeared synonymous with "white" (see "Second a Day" video from Save the Children, explored in greater detail later in this chapter).

4 Prosociality can be understood as a voluntary orientation toward the public versus private good (Baumeister and Vohs 2007).

5 See Upendra Baxi's *The Future of Human Rights* for a description of the dominant discourse of human rights as a neocolonial "gift" from "the West to the rest" (2006, 33).

6 Lanzoni (2018, 13) notes "that psychologists disagreed as to how to translate the German term 'Einfühlung' and 'empathy' ('in-feeling'), drawing on the Greek 'em' for 'in' and 'pathos' for 'feeling,' emerged as one among a number of alternatives."

7 See studies in communication and computer science that consider whether social media use is eroding our capacity to feel for and with others, all with little exploration of empathy's discursive significance beyond individual capacity (Alloway et al. 2014; Guan et al. 2019; Vossen and Valkenburg 2016).

8 Ai Weiwei has repeatedly drawn corollaries between his experiences as a political exile and those of people who have been victims of forced migration, despite his position as an internationally recognized artist putting him at a considerable economic remove. Although there are numerous political actions that use some of the same gestures or reenactments as those utilized by Ai Weiwei, this chapter is particularly interested in artworks that, because of the visibility of the artists creating them and their "viral" social media disseminations, have moved in circuits similar to international humanitarian-industrial models of awareness raising, rather than more immediate acts of "solidarity" such as those enacted by Moroccans on the beaches of Rabat (Linky 2015) who similarly "reenacted" the image of the death of Kurdi. Ticktin's conclusion in "Thinking beyond Humanitarian Borders" speaks to the political possibilities of the activist actions and ad hoc memorials at the US borders enacted by community members that treat mourning as a means of keeping "the dead in our world" as a call to action (Ticktin 2016). It is my contention that, by contrast, the large-scale international artworks projects I look at here, which memorialize "others," work to enclose and contain the dead and foreclose political possibilities of their presence in part because they were produced and circulated by and for people not identified with refugee communities.

9 See *Empathy*, the Stanford Art Center's exhibition at the Empathy Museum.

10 In February 2020, the MIA and CEVA became the most recent sites of Ai Weiwei's lifejacket installation.

11 An excellent example of how a memorial artwork that represents the death of refugees can be utilized toward right-wing arguments in favor of further illegalization of border crossing and greater policing (and militarization) of borders can be seen in the *Daily Mail* article "Touching or Creepy" about Finnish artist Pekka Jylhä's disturbing, realistic sculpture of Kurdi, "Until the Sea Shall Him Free." The article makes an almost immediate turn from the discussion of the artwork and the depiction of the child's death to a more protracted discussion of the trial of an alleged smuggler, making a none-too-subtle affective connection with the horror over Kurdi's death and the blame not of militarization of borders or the illegalization of migration, but firmly on the shoulders of criminalized migrant men, effectively making the boy's death an argument for greater security (Akbar 2016).

12 Late liberalism is both an era in time (beginning in the post-WW2 era and gaining ground after 1968) and a mode of governance related to but distinct from neoliberalism, in response to post-colonial social movements and anti-neoliberal resistance (25).

13 Povinelli articulates economies of abandonment as a tool of late liberalism used in the maintenance of sovereignty in which state violence is exerted not through overt or direct killing but rather through death as a result of the abandonment and neglect created by the divestment from social security characterizing neoliberalism (2011).

References

"7 Videos Guaranteed to Change the Way You See Refugees." 2015. UNHCR Innovation Service (blog). June 26, 2015. https://www.unhcr.org/innovation/7-videos-guaranteed-to-change-the-way-you-see-refugees/.

Agamben, Giorgio. 1995. *Homo Sacer: Sovereign Power and Bare Life.* Translated by Daniel Heller-Roazen. Palo Alto, California: Stanford University Press.

Ahmed, Sara. 2004. *Cultural Politics of Emotion.* Edinburgh: Edinburgh University Press.

AJ+. 2016a. *Ai Weiwei Won't Let You Forget about Europe's Refugee Crisis.* Youtube, 1:11. https://www.youtube.com/watch?v=7gAHIPYTGQE.

———. 2016b. *Amazing Underwater Museum Museo Atlantico.* Youtube. https://www.youtube.com/watch?v=H6oWA66TfqM.

Akbar, Jay. 2016. "Touching or Creepy? Artist Creates Sculpture of That Harrowing Image of Tragic Migrant Child Aylan Kurdi Lying Dead in the Sand." *Daily Mail* online. March 3, 2016. https://www.dailymail.co.uk/news/article-3475658/Artist-creates-moving-sculpture-tragic-migrant-child-Aylan-Kurdi-lying-dead-sand.html.

Alloway, Tracy, Rachel Runac, Mueez Qureshi, and George Kemp. 2014. "Is Facebook Linked to Selfishness? Investigating the Relationships among Social Media Use, Empathy, and Narcissism." *Social Networking* 3 (3):150–58. https://doi.org/10.4236/sn.2014.33020.

Archey, Karen. 2016. "For Photo Op, Ai Weiwei Poses as Dead Refugee Toddler from Iconic Image." *E-Flux.* February 2016. https://conversations.e-flux.com/t/for-photo-op-ai-weiwei-poses-as-dead-refugee-toddler-from-iconic-image/3169.

Arendt, Hannah. 1979. *The Origins of Totalitarianism.* San Diego, California: Harcourt Brace Jovanovich.

Arns, Inke. 2007. "Strategies of Reenactment." In *History Will Repeat Itself: Strategies of Re-Enactment in Contemporary (Media) Art and Performance,* edited by Gabriele Horn and Inke Arns, 37–63. Frankfurt am Main: Revolver.

Baumeister, R.F., and K.D. Vohs, eds. 2007. *Encyclopedia of Social Psychology.* SAGE Online Publications. https://doi.org/10.4135/9781412956253.

Baxi, Upendra. 2006. *The Future of Human Rights.* Oxford: Oxford University Press.

Berlant, Lauren. 2004. *Compassion: The Culture and Politics of an Emotion.* New York: Routledge.

Bernat, Roger. 2017. "The Place of the Thing." Performance. 28 April 2017. http://rogerbernat.info/en/shows/the-place-of-the-thing/.

Brown, Wendy. 2010. *Walled States, Waning Sovereignty.* Cambridge, MA: MIT Press.

Caldwell, Ellen C. 2018. "Can Art Help People Develop Empathy?" *JSTOR Daily.* January 16, 2018. https://daily.jstor.org/can-art-help-people-develop-empathy/.

"Center for Empathy and the Visual Arts." (n.d.). CEVA. Minneapolis, MN: Minneapolis Institute of Art. Accessed April 29, 2021. https://new.artsmia.org/empathy/.

De Genova, Nicholas. 2017. *The Borders of "Europe": Autonomy of Migration, Tactics of Bordering.* Durham, NC: Duke University Press.

Dhillon, Nitasha. 2016. "Ai Weiwei's Photo Reenacting a Child Refugee's Death Should Not Exist." *Hyperallergic.* February 3, 2016. http://hyperallergic.com/272881/ai-weiweis-photo-reenacting-a-child-refugees-death-should-not-exist/.

Donohoe, Janet. 2006. "Rushing to Memorialize." *Philosophy in the Contemporary World* 13 (1): 6–12. https://doi.org/10.5840/pcw20061311.

European Union Agency for Fundamental Rights (FRA). 2017. "Exploring the Connections between Arts and Human Rights." Report. Vienna, 29–30 May 2017. Luxembourg: Publications Office of the European Union. http://fra.europa.eu/en/publication/2017/exploring-connections-between-arts-and-human-rights-meeting-report.

Fassin, Didier. 2011. *Humanitarian Reason: A Moral History of the Present*. 1st ed. Berkeley: University of California Press.

The Global Citizen. 2016. "This is POWERFUL—Ai Weiwei Reflects Journey of Refugees with 14,000 Lifejackets in Berlin." *Twitter*. February 16, 2016. https://twitter.com/glblctzn/status/699775284980002816/photo/1.

Guan, Shu-Sha Angie, Sophia Hain, Jennifer Cabrera, and Andrea Rodarte. 2019. "Social Media Use and Empathy: A Mini Meta-Analysis." *Social Networking* 8 (4):147–57. https://doi.org/10.4236/sn.2019.84010.

Holmes, Seth M., and Heide Castañeda. 2016. "Representing the 'European Refugee Crisis' in Germany and Beyond: Deservingness and Difference, Life and Death." *American Ethnologist* 43 (1): 12–24. https://doi.org/10.1111/amet.12259.

Ivasiuc, Ana. 2018. "Sharing the Insecure Sensible: The Circulation of Images of Roma on Social Media." In *The Securitization of the Roma in Europe*, edited by Huub van Baar, Regina Kreide, and Ana Ivasiuc, 233–60. London: Springer.

Jayaraman, Gayatri. 2016. "Artist Ai Weiwei Poses as Aylan Kurdi for India Today Magazine." *India Today*. February 1, 2016. https://www.indiatoday.in/india/story/artist-ai-weiwei-poses-as-aylan-kurdi-for-india-today-magazine-306593–2016–02–01.

Lanzoni, Susan. 2018. *Empathy: A History*. New Haven and London: Yale University Press.

Le Guin, Ursula. 1976. "The Ones Who Walk Away from Omelas." In *The Wind's Twelve Quarters*, 275–84. New York: Harper and Row.

LGBTQIA+ Refugees Welcome. 2017. "Rockumenta." *Facebook*. May 21, 2017. https://www.facebook.com/lgbtqirefugeesingreece/videos/263576050783139.

Linfield, Susie. 2010. *The Cruel Radiance: Photography and Political Violence*. Chicago: University of Chicago Press.

Linky, C. 2015. "Group of People Reenact Drowned Syrian Toddler Image: Tribute or Gimmick?" *Catholic Online*. September 10, 2015. https://www.catholic.org/news/international/middle_east/story.php?id=63667.

Logan, Liz. (n.d.). "The Art of Empathy." *School of the Art Institute of Chicago*. Accessed January 18, 2021. https://www.saic.edu/news/marketing-communications/art-empathy-0.

Loughrey, Clarisse. 2016. "Ai Weiwei Made a Room Full of Celebrities Take Selfies in Refugee Jackets." *The Independent*. February 18, 2016. https://www.independent.co.uk/arts-entertainment/art/news/ai-weiwei-made-room-full-celebrities-take-selfies-refugee-jackets-a6881266.html.

Malkki, Liisa H. 1996. "Speechless Emissaries: Refugees, Humanitarianism, and Dehistoricization." *Cultural Anthropology* 11 (3): 377–404.

McCalman, Iain, and Paul A. Pickering, eds. 2010. *Historical Reenactment: From Realism to the Affective Turn*. London: Springer.

Mielczarek, Natalia. 2020. "The Dead Syrian Refugee Boy Goes Viral: Funerary Aylan Kurdi Memes as Tools of Mourning and Visual Reparation in Remix Culture." *Visual Communication* 19 (4): 506–30. https://doi.org/10.1177/1470357218797366.

Mirzoeff, Nicholas. 2011. *The Right to Look: A Counterhistory of Visuality.* Durham, NC: Duke University Press.

Moore, Deirdre. 2013. "Using the Arts to Build Empathy, the Ultimate 21st Century Skill." *The Institute for Arts Integration and STEAM* (blog). https://artsintegration.com/2013/05/15/using-the-arts-to-build-empathy-the-ultimate-21st-century-skill/.

Musiker, Cy. 2015. "Fighting the Empathy Deficit: How the Arts Can Make Us More Compassionate." *KQED*. September 3, 2015. https://www.kqed.org/arts/10933932/fighting-the-empathy-deficit-how-the-arts-can-make-us-more-compassionate.

Pedwell, Carolyn. 2012. "Economies of Empathy: Obama, Neoliberalism, and Social Justice." *Environment and Planning D: Society and Space* 30 (2): 280–97. https://doi.org/10.1068/d22710.

———. 2014a. *Affective Relations: The Transnational Politics of Empathy*. UK: Palgrave Macmillan. https://doi.org/10.1057/9781137275264.

———. 2014b. "Carolyn Pedwell on Empathy, Accuracy and Transnational Politics." *Theory, Culture & Society*. December 22, 2014. https://www.theoryculturesociety.org/blog/carolyn-pedwell-on-empathy-accuracy-and-transnational-politics.

———. 2016. "De-Colonising Empathy: Thinking Affect Transnationally." Edited by Sneja Gunew. *Samyukta: A Journal of Gender and Culture* 1, no. 1. https://kar.kent.ac.uk/54869/3/Pedwell_Decolonising%20Empathy%20Samyukta%20Jan%202016.pdf.

Perera, Suvendrini. 2009. *Australia and the Insular Imagination: Beaches, Borders, Boats and Bodies.* Ukraine: Palgrave Macmillan.

———. 2013. "Oceanic Corpo-Graphies, Refugee Bodies and the Making and Unmaking of Waters." *Feminist Review* 103:58–79.

Povinelli, Elizabeth A. 2011. *Economies of Abandonment: Social Belonging and Endurance in Late Liberalism.* Durham, NC and London: Duke University Press. https://doi.org/10.1215/9780822394570.

Rajaram, Prem Kumar. 2002. "Humanitarianism and Representations of the Refugee." *Journal of Refugee Studies* 15 (3): 247–64.

Save the Children. 2016. "Most Shocking Second a Day Video." Youtube, 1:33. https://www.youtube.com/watch?v=RBQ-IoHfimQ&t=78s.

Slovic, Paul, Daniel Västfjäll, Arvid Erlandsson, and Robin Gregory. 2017. "Iconic Photographs and the Ebb and Flow of Empathic Response to Humanitarian Disasters." *Proceedings of the National Academy of Sciences* 114 (4): 640–44. https://doi.org/10.1073/pnas.1613977114.

Smillie, Susan. 2016. "Drowned World: Welcome to Europe's First Undersea Sculpture Museum." *The Guardian.* February 2, 2016. http://www.theguardian.com/artanddesign/2016/feb/02/drowned-world-europe-first-undersea-sculpture-museum-lanzarote-jason-decaires-taylor.

Stanford University Gallery. 2015. "Empathy." *Lecture.* August 12, 2015. https://arts.stanford.edu/event/empathy/.

Taylor, Jason deCaires. 2015. "Transcript of 'An Underwater Art Museum, Teeming with Life.'" TED video, 11:00. Posted December 2015. Accessed August 30, 2021.

https://www.ted.com/talks/jason_decaires_taylor_an_underwater_art_museum_teeming_with_life/transcript.
Ticktin, Miriam. 2011. "The Gendered Human of Humanitarianism: Medicalising and Politicising Sexual Violence." *Gender & History* 23 (2): 250–65. https://doi.org/10.1111/j.1468-0424.2011.01637.x.
———. 2016. "Thinking beyond Humanitarian Borders." In *Social Research: An International Quarterly, Borders and the Politics of Mourning* 83 (2): 255–71.
Vossen, Helen G.M., and Patti M. Valkenburg. 2016. "Do Social Media Foster or Curtail Adolescents' Empathy? A Longitudinal Study." *Computers in Human Behavior* 63: 118–24. https://doi.org/10.1016/j.chb.2016.05.040.

PART TWO

STATE POLITICS AND GLOBAL GOVERNANCE

4

A Means to a Shelter

Sally Souraya

Waiting for the car to pick me up after a field visit to a refugee camp for Syrians in Lebanon, the time felt long, the heat intensified, and the dust was constantly blowing in my eyes. For a moment, everything seemed blurry as I was trying to replay in my mind the events of the whole day. I looked around once again, trying to absorb the reality of the place and its people, and suddenly, the word *hudud* (حُدود) in Arabic, which means "borders," stood out, perhaps to sum up a human tragedy represented in one place.

The word was on a tent I was standing near, on an old banner laid across a fading tarp from which the tent was made. Ironically enough, the original line was actually "لا حُدود" (No Borders), yet the tarp was folded over in a way that hid the "No." What was mostly just an incidental detail in the assembly and use of the tent turned out to be an eye-opening, meaningful coincidence for me, as I quickly took a photo of the tent and exchanged a few words with members of the family who were living there. Watching the family moving in and out under what was left of that obscured slogan made me think about the paradox created here. Being just around ten miles away from the Syrian border, returning to cross the borders for them as refugees was not even an option.

I left the camp on that day, but since that moment in 2017, this memory has stayed with me. I knew there was more to see, explore, and tell about the refugee crisis from the perspective of visual arts. In 2018 and 2019, I went back to Lebanon and visited more camps in the Biqa' region, and then I started working on this photo series titled *A Means to a Shelter*. As an ongoing project,

Fig. 4.1 لا حُدود / لا حُدود Borders\No Borders, *Photo by Sally Souraya, Lebanon, 2017.*

Fig. 4.2 من From, *Photo by Sally Souraya, Lebanon, 2019.*

Fig. 4.3 صنع في المانيا Made in Germany, *Photo by Sally Souraya, Lebanon, 2019.*

the series explores how leftover or donated advertising tarps used for makeshift tents in Syrian refugee camps could represent alternative narratives for advocacy. Beyond the practicality of the tents, I wanted the photos to highlight how the advertisements juxtapose ironically with the temporary life of people in the camps and the feeling of "home," creating almost coincidental graffiti.

Throughout this project, I was interested in understanding the linguistic landscape and visually interpreting the stories that these slogans/banners revealed, as they blended within the refugees' context. I did not want the photos to just document what remained of these advertisements on the tarps, their altered or new meanings, but also to give the viewers a way to reflect on what these simple yet powerful messages mean to the people who live literally under or behind these words.

With words being the center of this photo series, the idea is to understand how language helps in eliciting the realities of refugee conditions but also in reviving hopes among people. Could language itself become a shelter and a refuge for people who live in the camps? Focusing mainly on tents made of Arabic advertisement tarps, the photos try to question how refugees identify with and relate to these words/messages that surround them. Could these messages become their narrative/voices to assert their current state of living or perhaps to form a basis for protest? Could they offer them signs of belonging and resilience or anchors to dreams of better homes and better lives?

The photos try to emphasize the patchwork that the tarps create on the tents. There is something surreal not only in the meanings that the words hold but also in the way these words are laid out to make the tents. In a way, while the layout could purely be inadvertent, some of the messages on the tents feel as if they have, somehow, been curated by people living in the camps and those working in the humanitarian sector.

As the project evolved from one tent to another, it became clearer to me that the focus of each image should be on the words to speak for themselves. Therefore, I tried, where possible, to take close-up photos excluding most of the surrounding context of the tents. Equally, I chose not to include people to preserve a sense of anonymity in the photos and to respect people's privacy.

Throughout the process of developing the photo series, from talking with people in the camps to taking photos, I intended to keep my artistic approach more intuitive and let the project evolve organically. In a way, I did not want to be prescriptive or force the process by searching for tents that would "fit" within my project. I was just allowing myself to be there in the camps, connecting with the place and people, ready to capture whatever I would encounter in relation to the project theme.

This photo series is more than just an artistic project that tells the story of a place. It highlights how the tents and living conditions of people in refugee camps go beyond being just "a means to a shelter." At the end of each of my visits to the camps, while waiting for the car before leaving, I would look again at the tents and witness how some of the messages on the tarps were fading with time. I then asked myself: Will *they* survive the next winter?

5

Managing Crisis? (Re)Thinking Turkey's Discourse of "Refugee Crisis"

Eda Sevinin

Introduction

In late 2018, during a conference in Istanbul that focused on comparing the migration and citizenship regimes of France and Turkey, I had the chance to listen to some of the presentations. One of them, given by a well-known migration scholar in Turkey, focused on the differences in responses to migration in Turkey and Western Europe. He suggested that in Turkey, "we"—the migration scholars, NGOs, and the state institutions—should start addressing the current situation of migration as a "crisis." He further claimed that only after Turkey starts naming what is going on as a "crisis" would it be able to "seriously" respond to migration in terms of policymaking, integration, and social cohesion. For him, the main problem was the indifference regarding refugee management, which, in turn, led to ineffectiveness in policymaking and integration policies. Taking the situation seriously, in fact, calling it an "emergency," a "crisis," could be the way out of this deadlock and uncertainty that marks the asylum system in Turkey.

On the one hand, it was true that Turkey was not referring to the presence of four million refugees (UNHCR 2019) as a "crisis." It was also undoubtedly true that Turkey's asylum system was identified with ambiguity and uncertainty (Biehl 2015) that, in a way, immobilize refugees and asylum seekers in an eternally temporary status, which is sometimes called being "in limbo."[1] On the other hand, the abovementioned point calls for further questions: What does it mean for Turkey to eschew calling it a "crisis," and how is it to be explained? Given Turkey's "transit" location, how was it implicated in the rhetoric of "crisis" that has swept the EU since 2015? What kind of resources could be mobilized in the asylum regime if the situation was addressed as a "crisis," and for whom?

This chapter is an attempt to think through these questions. I investigate how Turkey approached the "migrant/refugee crisis"[2] discourse and the political implications of this approach in terms of how Turkey (re-) locates itself in the global geopolitics. I will look at media and the political rhetoric and argue that in both, "migrant/refugee crisis" was attributed to the failure of the Western countries, predominantly of the EU countries, to live up to the promises of certain ideals—humanity, human rights, and conscience—to which they subscribe. In doing so, the Turkish nation-state (with all its institutions and representatives, including the media and parts of civil society) postulates a moral primacy over the Western imaginary of human rights.

This postulation has multiple simultaneous functions. First, it legitimizes and justifies Turkey's migration and asylum management based on "humanitarian government" (Fassin 2011) of refugees instead of fully recognized rights and status. Second, while it creates a spectacle of welcoming Syrian refugees as a sign of Turkey's ability to manage "crisis," it simultaneously invisibilizes other refugee groups residing in Turkey. Thus, Turkey's asylum regime is grounded on the contradiction between the language of rights and charity/hospitality. This, in turn, enables more critical governmental interventions by producing a form of refugeehood that is at the disposal of the governmental powers.[3] This contradiction can also be specified as "the tension between the humanitarian and the political production of the asylum seeker/refugee in the contemporary refugee management" (Rozakou 2012, 563). Humanitarian production of the refugee, as widely discussed in the literature (Fassin 2011; Malkki 1996; Ticktin 2011; Rozakou 2012), allows for biopolitical interventions that can range across a spectrum between to "make live" and to "let die" (Li 2010), or, in Turkey's case, "make stay" and "let go." As we have witnessed since the readmission agreement signed in 2016, colloquially known as the EU-Turkey deal, these interventions can amount to immobilization[4] as well as forced displacement (as in the recent deportations, which started in the summer of 2019) of people in a given territory. Finally, such an understanding crystallizes and materializes imagined geography in a reverse order: on the one hand, the West, failing and reluctant to respond to a humanitarian crisis; on the other hand, Turkey, claiming supremacy over the West by assuming and re-claiming putatively non-Western moral traits of hospitality, conscience, and commitment to religious duties.

In the first section of the chapter, I will briefly outline the discourse on the "migrant/refugee crisis" that has been widely circulated in Europe. I will touch upon what the crisis discourse has enabled and disabled politically. In the second section, which is also the main section of the chapter, I will focus on the dominant discourses of "crisis" in Turkey.

The "Crisis of Europe"

Recently, population movements at the borders of Western countries, and most notably, the European Union, have been specified and addressed as a "migrant and/or refugee crisis" by policymakers, NGOs, international organizations, and the media. Although the discourse on "crisis" is hardly new in the EU's migration and asylum regime (see Bojadžijev and Mezzadra 2015), the population movements that started in late 2014 and continued throughout 2015 dominated the ways in which migration and asylum were previously discussed, understood, conceptualized and, finally, politically addressed. It can be argued that "crisis" discourse has become virtually the only paradigm for discussing the issue of "migration" in and to Europe (Cantat 2016).

Population movements in different forms have constituted an essential part of the history of the West in the foundation of the nation-state and the consolidation and progress of capitalism (Marfleet 2016). Here, one can think of guest-worker agreements, population movements from former colonies to the colonial metropoles, Cold War refugees who were welcomed in the West as "freedom fighters" against the communist regimes, the war in the former Yugoslavia, as well as historical forms of forced displacements and enslavements in colonial history. In this context, the question becomes why is it a "crisis" at this historical juncture, and what does it mean for the politics of mobility?

Bojadžijev and Mezzadra (2015) start their article with the question, "How did the 'crisis' begin?" For them, the origins of the crisis are, undoubtedly, not recent and go farther back than 2015. They remind us of other moments that turned the "Mediterranean into one of the most dangerous and lethal borders of the planet" in the 1990s. Intensification of migration from Albania in 1991 and Tunisia following the "Jasmine Revolution" in 2011 had already called for "emergency measures" for "the reorganization, tightening, and even militarization of the European maritime border regime" (Bojadžijev and Mezzadra 2015). Thus, events that the world witnessed in 2015 should be located in a longer history, albeit with qualitative and quantitative differences. Migration policies that transformed the Mediterranean into one of the deadliest borders in the world sparked intensive debate about the crisis-like situation going on at the EU peripheries. Nevertheless, how did the crisis of 2015 start?

As Céline Cantat (2016) states, the discourse of crisis gained currency with a series of four consecutive shipwrecks, which caused the death of more than twelve hundred people in April 2015. In the following months, increasing death tolls in the Mediterranean became the center of attention for European media and political discourse. On September 3, 2015, the photo of three-year-old Alan Kurdi's body on the shores of Turkey marked the first significant event (Cantat 2016). The march of refugees from Budapest to Vienna—referred to as the "March of Hope"—marked the second event in which the "real crisis" began (Bojadžijev and Mezzadra 2015; Kallius, Monterescu, and Rajaram 2016; Hess and Kasparek 2017). Immediately after, German Chancellor Angela Merkel announced the unilateral suspension of the Dublin II Regulation, which stipulates the "first-country-of-arrival" rule and opened the doors for refugee arrival in Germany. Although this first moment of opening was followed by "multiple closures" (Bojadžijev and Mezzadra 2015), Germany's move, with even the *temporary* suspension of the Schengen Agreement—the epitome of intra-EU freedom of mobility—was largely regarded as the announcement of the need for extraordinary measures both within the EU and in the neighboring countries.

The use of emergency measures, some of which are still in place, brought along the dissemination of the crisis discourse. This facilitated various political interventions, including tightening the asylum laws in the European countries, reimplementing and even strengthening the border controls in the Schengen Zone, militarizing the external borders of the EU, erecting walls and razor-wires at the EU borders, and accelerating readmission agreements with the neighboring countries such as Turkey, Libya, and Serbia as part of the "externalization of the border" scheme. These measures had several political consequences, not only for the people on the move but also for the EU and the nation-states involved in the process.

Amid the debates and policy discussions uncritically embracing the "crisis" in the fall of 2015 and the following months, we also heard critical voices from scholars of critical migration and border studies and migrant solidarity networks organized by, for, and with migrants. This group contended that what has been named and framed as a "migrant/refugee crisis" is the crisis of nation-state borders and, more particularly, "the crisis of the EU as a political project and its border politics" (Cantat 2016; also see Rajaram 2015; New Keywords Collective 2016). That being said, the broader question is what the crisis framework enables, reveals, or conceals.

It can be argued that the pace and scale of migration from and to various geographies of the world has increased and, further, has created a situation requiring emergency measures. On the other hand, coupled with the crisis discourse, we also witness the reproduction of Western-centric representations of geography. Mainstream accounts of migration scholarship, policymakers, and media have displaced the focus of migration through discursive and political means. We witness a representation of unidirectional human mobility *to* the West *from* countries of the Global South despite the statistical data that show otherwise. In turn, this reproduces and strengthens the West's historically, politically, and discursively constructed supremacy over non-Western geographies while simultaneously reproducing a politics of geography built upon the West/non-West binarism.

In addition, the "refugee/migrant crisis" framed the "unsanctioned" human mobility as abnormal and illegitimate. As Nicholas De Genova (2013) argues, migrant illegality is normalized and naturalized through border laws and regulations. The nation-state and its borders are historical constructs, and migrant illegality is determined through the border- and law-making authority, in its current form, the nation-state. Any forms of human mobility that challenge or contrast with the borders defined by the political power are thus classified as illegitimate, illegal, and extraordinary measures vis-à-vis the cross-border mobilities that are regarded as justifiable and necessary (Nyers 1998). The crisis discourse constructs certain people on the move as "illegal and illegitimate migrants" whose mobility should be controlled, constrained, and halted through various means of physical, economic, and sociopolitical immobilization. This strategy of constructing legal and illegal migrants as well as "genuine" and "bogus" refugees enables (bio)political interventions directed at a particular population.

Moreover, the temporality of "crisis" that cannot afford long-term political debates and negotiations (Nyers 1998) closes the political space for many actors, more particularly, for the migrants themselves. Thus, the refugees' and migrants' access to a political space in which they are the most implicated is thwarted as yet another form of immobilization. These constraints for the migrants and refugees are transformed into the further political and physical mobility of other actors who are able to enact, strengthen, and reproduce the crisis framework. While in the Hungarian border spectacle, this is demonstrated by the armed forces deployed at the borders (Cantat 2020), in Turkey, it is demonstrated by the humanitarian organizations, NGOs, and international organizations.

The "crisis" framework is a political project that serves not only the restoration of a "national order of things" (Malkki 1996) but also the consolidation of "legitimate actors of mobility." Naturalizing nation-states and (more often than not violently) emplacing human societies to artificially delineated territories render people on the move, particularly refugees, anomalies to the "order of things." Rendering mobile non-Western peoples as an "anomaly" functions as a dehistoricizing and depoliticizing force in accounting for human mobility (Berg and Fiddian-Qasmiyeh 2018). By dehistoricization, I refer to two things: first, dehistoricizing inspirations, aspirations, imaginations, desires, affects, and mobilizations of people on the move; second, dehistoricizing the implicatedness of Europe as a colonial space (Roy and De Genova 2018) in the current affairs that are labeled as a "crisis" by the West. Even if not a singular historical actor itself, the West is very much implicated in the racialized hierarchies among nations produced through colonial histories and white supremacy (De Genova 2018). Non-Western people on the move are seen through this historical filter and positioned within these lingering histories, which condition the international order as well as asylum regimes of nation-states. Asylum regimes of the West are, thus, racialized orders where some nationalities (according to racialized colonial histories) were rendered less mobile, less skillful, and less acceptable by "Western standards" (Czajka 2015). Migration from the Global South to the Global North is conditioned not only by nation-state-centric border management but also by the representation and construction of racialized subjects of the so-called Global South.

The "Migrant/Refugee Crisis" in Turkish Discourse

While the crisis discourse was at the center of EU migration and asylum discussions, Turkey's position vis-à-vis forced migration and the "migrant/refugee crisis" was slightly different. Before moving to Turkey's responses to the so-called crisis, I will give a brief account of Turkey's asylum system.

Turkey's Asylum System: A Brief Overview

Recently, Turkey has been a focus of the UNHCR agenda as the country hosting the largest population of refugees worldwide. According to UNHCR figures, as of May 2020, Turkey is hosting almost 4 million refugees and asylum seekers, 3.6 million of whom are people who came from Syria (UNHCR 2019). Although these 4 million are usually referred to as "refugees and asylum seekers," legal statuses in Turkey are more complicated. As one of the original parties to the Convention in 1951, Turkey implemented a geographical limitation to the Geneva Convention and its 1967 Protocol. The geographical limitation purports to be upholding the clause "events occurred in Europe." This, in turn, allows Turkey to keep asylum and refugee status applicants from non-European countries of origin in limbo without a clearly recognized status.

In Turkey, a comprehensive migration law was absent until very recently (Soykan 2010, 6). In April 2013, due to the EU accession process and many Syrians coming to Turkey, a new law, the Law on Foreigners and International Protection (LFIP), was adopted. It became the first comprehensive legal regulation of the immigration system in Turkey (Kilberg 2014). Different categories of asylum have been recognized with this law, yet the geographical limitation has not been lifted. Individual non-European asylum seekers are recognized as "conditional refugees" waiting for third-country resettlement. Moreover, another status, "subsidiary protection," is for people "who neither could be qualified as a refugee nor as a conditional refugee" and who "shall nevertheless be granted subsidiary protection upon the status determination because if returned to the country of origin or country of [former] habitual residence"[5] (Law No. 6458, 2013) would face the threat of death sentences, torture, or ill-treatment.

In 2014, another regulation called the Directive of Regulation on Temporary Protection for those who flee their country of origin *en masse* and cannot be returned was adopted (Rygiel, Baban, and Ilcan 2016). The LFIP and the temporary protection regulation clearly stipulated that "status of temporary protection provides an indefinite temporary residency for Syrians [as well as refugees and stateless people from Syria] fleeing the civil war without a long-term promise of residency or citizenship rights or even the ability to claim refugee status" (Rygiel, Baban, and Ilcan 2016, 317). Although the law that regulates immigration was eventually codified in Turkey, the legal framework hardly meets the social, economic, and political requirements of the practice (Cavdar 2016; Özden 2013; Ataç et al. 2017). With such a system, Turkey's migration regime is characterized as a multilayered (Genç, Heck, and Hess 2018) and fragmented system with multiple actors in the field: governmental authorities, United Nations, third parties, and local and international NGOs (Sarı and Dinçer 2017).

Scholars focusing on Turkey's asylum regime argue that, since the 1990s, Turkey's asylum regime has been shaped in accordance with EU migration and border policies (Genç, Heck, and Hess 2018; İçduygu 2011; Özçürümez and Şenses 2011). Genç, Heck, and Hess (2018, 490) show that as early as 2002, Turkey had become "a target for EU-migration control." Harmonization efforts notwithstanding, this process was full of controversies, contradictions, and obstacles (Genç, Heck, and Hess 2018). The most enduring contentious issue pertains to Turkey's insistence on upholding the "geographical reservation." While the EU system wants Turkey to lift the geographical reservation as a condition of aligning its migration and asylum regime with that of the EU, the Turkish state has steadfastly refused.[6] Besides these obstacles within the harmonization process, changes in the foreign (and domestic) policy trajectory and transformations in approaches to "Europeanization and Westernization" during consecutive AKP governments had immediate effects on migration and asylum systems. As Sibel Karadağ (2019, 2) puts it, "Turkey presents an interesting case – a candidate country whose ambitious foreign policy is moving away from Europe with the motivation of being a 'regional actor' in the Middle East under increasingly authoritarian rule."

Turkey's "Migrant/Refugee Crisis"?

As briefly outlined above, Turkey's asylum regime bears contradictions and ambiguities not only regarding its responses to the European border and asylum regimes but also in terms of its responses to refugees. In other words, Turkey's asylum regime and associated official discourses are not linear but contradictory, ambiguous, and uncertain. This can be observed in the role Turkey assumes in the "migrant/refugee crisis" framework. On the one hand, it developed welcoming public rhetoric, exclusively directed toward the Syrian refugee situation; on the other hand, it continued militarizing its Greek and Bulgarian borders, erecting a 564-kilometer concrete and heavily militarized wall on the Syrian border, and implementing violent deportation schemes. The most obvious moment was the EU-Turkey readmission agreement (widely known as the EU-Turkey deal) signed in March 2016 (Genç, Heck, and Hess 2018; Rygiel, Baban, and Ilcan 2016; Soykan 2010; Heck and Hess 2017). However, instead of seeing these moments as mutually exclusive, I argue that Turkey's entanglement with the "crisis" framework enables Turkey to deploy securitized (and, at times, militarized) measures toward refugees while continuing the welcoming and humanitarian discourses and practices.

The arrival of refugees in "unprecedented numbers" to Turkey began before 2015 when Europe experienced its infamous "summer of migration" (Ataç et al. 2017; Rygiel, Baban, and Ilcan 2016). Turkey's approach to refugees was found laudable by many international organizations, especially that of the open-door policy implemented between 2011 and 2015 (Ataç et al. 2017). While welcoming discourses toward Syrian refugees was the government's official rhetoric, Turkey also distinguished itself from the West; it developed a distinct language of "crisis management." It should be noted here that I am not suggesting Turkey has managed the "crisis." What I am discussing is that Turkey developed a nation-state identity that, in turn, implicated itself in the "solution" of what has been posed as a problem. The notorious readmission agreement, the recent deportations of undocumented Syrian refugees back to Syria, and the military and diplomatic operation to create a "buffer zone" in northern Syria to "send the refugees back" are fundamental to this self-proclaimed role of crisis management.

This "crisis management" language provided a strong position for Turkey vis-à-vis the West, which has been erecting walls and implementing strengthened security policies against refugees. That is, Turkey found a discursive space to construct a moral high ground to criticize the West. As shown in the then Prime Minister Erdoğan's words:

> We opened our doors to our siblings who were fleeing from conflicts in Iraq and Syria and mobilized our means. We now host more than 1.5 million people in our country. Why? This is our understanding of *humanity, conscience, and Islam*. That's why we did it. We could not leave them to the danger of terrorist acts, bullets, bombs. We could not leave them to the murderous Assad regime. If they emigrated to this country, we were obliged to be an *Ansar*. And we did it. And we still do . . . At present there are only 130,000 asylum seekers in Europe, and Europe complains about it. But only in Turkey, there are 1.5 million asylum seekers. This is our *difference* compared to the West (Erdoğan 2014 quoted in Öztürk 2017, emphasis added).

This official discourse, mostly voiced by the then Prime Minister and now President Recep Tayyip Erdoğan, seems to be informed by the language of humanitarianism and hospitality, which he summarized as "humanity, conscience, and Islam." In a similar vein, migration scholars from Turkey criticized European states for turning their backs on refugees prior to the onset of migration toward European borders in 2015 (Akdoğan 2018; İktisadi Kalkınma Vakfı 2016; Şen and Özkorul 2016). Mevlüt Çavuşoğlu, the minister of foreign affairs since 2014, wrote a piece in 2016 titled "Turkey and the Syrian Refugee Crisis: An Example for Humanity," in which he discusses Turkey's role in this "humanitarian catastrophe" as opposed to how Europe reacted to this process:

> The global response to this humanitarian catastrophe has not been quick enough or comprehensive enough to help alleviate the suffering.

> The EU countries are deeply divided on the refugee crisis and lack a concerted and coordinated approach to their migration and asylum policies. As a response to the crisis, several countries introduced more restrictive immigration policies and enacted harsh measures against migrants. Entry denials and turning away refugees at the borders of some European countries are nothing less than embarrassing (Çavuşoğlu 2016, 18).

I would like to discuss what this discourse reveals and conceals and what might be the political consequences. Importantly, although these discourses that underline and criticize Europe's failure in terms of refugee reception have persisted, it is difficult to argue that they are linear or perfectly overlap with Turkey's practical responses. Turkey's position in these discourses and the crisis framework is politically and contextually different from that of the EU. The first reason is historical and contextual. Given that the arrival of many refugee populations in Turkey started long before 2015, refugee management in Turkey has developed a different course. The second reason is political: Turkey's approach to refugees (especially the Syrian refugees) was discursively and practically—albeit limited—based on a humanitarian and (Sunni) Islamic language that is informed by the reified culturalist and nostalgic accounts of hospitality under Ottoman rule before the twentieth century. Nonetheless, the legal status of migrants remains a critical issue since Turkey, despite hosting the largest number of migrant populations, does not have a comprehensive asylum system but rather a fragmented system with multiple actors operating simultaneously (Sarı and Dinçer 2017).

Due to Turkey's geographical location, the Turkish asylum regime after 2015 has become an indispensable part of the language of "migrant and refugee crisis." The Turkish state has never challenged the language of crisis, yet it has developed a relatively distinct language of "crisis management." Islamic language has been useful in the development and deployment of this discourse. "Humanitarianism," which can be framed as the moral command for the "care of the other who is in need," is entangled with the Islamic and cultural command of "hospitality," the narrative of *ansar* and *muhajirun*.[7]

This language of crisis management allowed Turkey to frame the "migration crisis" as solvable through "appropriate means" and by "appropriate actors." While Turkey capitalized on the potential political consequences of the crisis discourse in terms of centralizing and monopolizing the political space, as well as deploying culturalist discourses (Rajaram 2015), it also eschewed calling it a "crisis" of Turkey. Instead, Turkey posed itself as the problem solver both internally and externally. In what follows, I will discuss the political consequences of this "crisis management language." In doing so, I will focus on two central tensions enabled by Turkey's implementation of crisis discourse. The first is the concealment of the discrepancy between languages of rights and languages of charity/hospitality, which can also be specified as the tension between "the humanitarian and political production of the asylum seeker/refugee" (Rozakou 2012, 563). The second one pertains to how the geographical imagination of Turkey is revealed through its approach to the crisis framework.

As mentioned above, an uncertain and limited legal framework has many consequences. Geographical limitations and the involvement of multiple actors posited a gap in migration management. On the one hand, there is the legal framework that limits refugee rights through ambiguous statuses. On the other hand, a large space was opened for the humanitarian interventions undertaken by the local, national, and international NGOs and state institutions. This humanitarianism became the main tenet of refugee management in a way to conceal the exclusion of refugees from basic rights, as well as hierarchizing various refugee groups in accordance with their conformity to Turkey's Islamically oriented humanitarian logic. With the humanitarian way of governing, refugees were constructed as the objects of humanitarianism who are open to a spectrum of biopolitical interventions (Li 2010). Also, this humanitarian way of governing easily erased refugees' political subjectivities as well as aspirations, desires, hopes, and frustrations and dehistoricized refugee movements. Thus, in the spectrum of humanitarian and political production of refugeehood, Turkey located the refugees at the humanitarian end thanks to its capacity to cast itself as powerful enough to manage (and, at times, exacerbate)[8] the "crisis." This being the case, through humanitarian language, the Turkish nation-state manages to conceal this discrepancy and poses itself as the protector. Here, protection serves two meanings: on the one hand, Turkey protects refugees from dire conditions back in their countries. On the other hand, Turkey protects Europe as an imagined destination of refugee movements from "refugee flows."

Related to the second category, this form of refugee management, "Turkey's way," as it were, has repercussions regarding the political geography of Turkey. In Turkey, the "migrant/refugee crisis" was immediately attributed to the West, more particularly to the EU, which, in turn, helps to externalize Turkey's role in the refugee management regime of Europe. The crisis was framed as the "crisis of Europe," and the Turkish state has constantly reiterated the EU's dependency on Turkey for crisis management. Juxtaposing itself vis-à-vis the West (more particularly, the West as an imagined geography), Turkey posed itself as the country that is welcoming refugees as opposed to the West, which erected and militarized its borders (see the quotations above). One well-known migration scholar in Turkey joined these efforts and asserted that what the EU was going through was "not a migrant crisis but rather a crisis of humanity" (Erdoğan 2016). Both in Murat Erdoğan's (2016) and state officials' accounts, the EU is held accountable for not embracing values of humanity and human rights, which have been the foundational premises of the EU. Both accounts also accuse the EU of reluctance to take up its share of the "burden" of refugees, both financially and human-wise. Erdoğan continues: "The EU is one of the richest regions in the world, holding 31% of the total GDP of the world . . . Despite all this wealth and welfare, it still maintains its objection to refugees" (Erdoğan 2016). In his comparison between Turkey and the EU, Erdoğan (2016) states that "Turkey, despite the scarce resources it has, has been hosting the largest refugee population whereas the EU, albeit doing much less, keeps pointing out the human rights violations in Turkey which opens its doors to refugees with 'humanitarian concerns.'"

However, this comparison and geographical relocation of Turkey does not necessarily challenge the politically central position of the West—as seen in the readmission agreement. It rather denotes a paradoxical moment for Turkey. This reorientation, I think, serves two functions simultaneously: first, it works to contain and conceal Turkey's role as Europe's "gatekeeper," a role that is also funded by the EU. Secondly, it helps Turkey to reinforce itself as morally superior vis-à-vis the West, which, in turn, allows it to deploy a discourse that works both internally and externally.

In conclusion, my argument is that this juxtaposition of the humanitarian discourse vis-à-vis the "crisis" discourse is a policy of containing political discussion on refugee management in Turkey and beyond. In Turkey's current policy, humanitarianism and "crisis discourse" are not mutually exclusive but rather mutually enabling. Insofar as the migratory movements are framed as a "crisis" to pave the way for and accelerate emergency measures, humanitarianism and humanitarian discourse are posed as the "solution" to the crisis, albeit temporarily. However, as a nation-state policy and as an emergency measure, humanitarianism does not exclude other emergency measures that the political power sees fit. In other words, "make live" and "let die" (or "make stay" and "let go") interventions enabled by the ostensible dichotomy between border policing by the EU and humanitarianism by Turkey, respectively, are not mutually exclusive but rather mutually reinforcing in a way to accord more power and authority over the lives of refugees and migrants who were immobilized within the borders of Turkey.

Notes

1 In Turkey, "refugee" as a legal status is granted only to people who conform to the geographical scope in which Turkey accepted the 1951 Geneva Convention. Turkey upholds a clause that the convention only applies in relation to "events occurred in Europe" ("Europe" applies to Council of Europe member states). Accordingly, people who came from "non-European" countries are granted "conditional refugee," "subsidiary refugee," or "temporary protection" statuses. Legal regulations notwithstanding, throughout the chapter, I will not make a distinction between legal categories and use the term "refugee" in order to challenge the legal categorizations of people who have been displaced and who are entitled to claim international protection. The distinctions and legal implementation of these categories are beyond the scope of this chapter. For more information, see Sarı and Dinçer's 2017 paper "Toward a New Asylum Regime in Turkey?"

2 Framing crisis as "migrant crisis" or "refugee crisis" are qualitatively and politically distinct discourses. "Migrant crisis" is often based on the differentiation between refugees and "economic migrants" and casts doubt on the type of urgent needs implicated in mobility, thus as merely "economic," while the "refugee crisis" framework is based on humanitarian government of asylum regimes in a way to cast refugees as "victims of wars and disasters" in which Europe (and the West, in general) can externalize itself from the "humanitarian tragedies" that have happened in "far away geographies." For a detailed analysis of this differentiation, see Cantat (2016) and Apostolova (2015). Drawing on this line of thought, throughout the chapter I use "migrant/refugee crisis."

3 Here, I use "governmental powers" in order to avoid falling into the trap of an analysis that prioritizes the nation-state as the only actor in refugee management. I refer to state and non-state actors who have become a part of the "crisis-management discourse" in Turkey. The discourse might seem embodied by President Erdoğan's leadership; however, discourses and practices of crisis management encompass a much larger group of actors.

4 Apart from registered Syrians who are under the category of "temporary protection," refugee populations in Turkey have long been subjected to politics of immobilization through various technologies such as satellite cities (see Biner 2014), issuing (or rejecting to issue) travel permits to change cities. That is, "once relocated, refugees have to sign in regularly at the DGMM (Directorate General of Migration Management) and cannot leave their city without travel permits issued by the DGMM. The main purpose of travel permits and regular check-ins is to regulate and control refugees' movements" (Sarı and Dinçer 2017). However, throughout late 2017 and early 2018, "Syrians who were already registered in a certain district would now have to apply for official travel permits in order to leave that district, turning even small family visits and business trips into a bureaucratic nightmare" (Netherlands Helsinki Committee 2019).

5 Law on Foreigners and International Protection No. 6458. Available at https://en.goc.gov.tr/kurumlar/en.goc/Ingilizce-kanun/Law-on-Foreigners-and-International-Protection.pdf.

6 The reasons provided for Turkey's insistence on the geographical limitation are manifold. Officially stated reasons usually underline the geographical proximity to

the refugee-generating regions such as the Middle East and Africa. The geographical location of Turkey and the political situation to its east are still used as the main reasons behind the lack of rights and status. However, the unspoken factors underlying nationalist concerns also include, first, population engineering as to who will be included within the territory that was historically Turkified and, second, the complex relations that arise from its position between the Middle East and Europe.

7 Dawn Chatty (2010), for instance, has a different position on the issue. She historicizes hospitality as a cultural trait that can be traced back to the late Ottoman period. She argues that in the former Ottoman territories that are now called "the Middle East," hospitality (guest and host relations) was a response to the problem of displaced populations. I appreciate her attempts at tracing the historicity of a seemingly nation-state discourse and policy. She points out more community-based responses to stateless people and challenges the prioritization of nation-state-centric responses to refugee reception. However, I believe this account should be approached carefully for two reasons. First, introducing somewhat reified cultural and religious traits into a right-based sphere, that is, forced displacement, jeopardizes legitimating an asymmetrical relationship between the newcomers and the host, and it places the communitarian (and cultural) values at the center of refugee reception. Secondly, hospitality in the Ottoman context also had its darker side, systematically displacing and dispossessing non-Muslim communities under the political operations of "population engineering" (Dündar 2015; Watenpaugh 2015). This being the case, I think hospitality discourses conceal the historical responses to mass-scale displacement in the Ottoman period as much as they reveal them.

8 At the time of writing, the migration agenda in Turkey drastically changed: on February 27, 2020, upon an attack on Turkish Armed Forces in Idlib, Syria, which killed more than 30 Turkish soldiers, the government unilaterally annulled the Turkish-EU Readmission Agreement and announced that Turkey would open the European borders to refugees. The next day, hundreds of refugees set off to the Greek border and attempted to continue their journey toward the EU. According to the Minister of Interior Süleyman Soylu, 142,145 people crossed the Greek border. However, those who went to the land borders (approximately 5,000 people) were stranded in the border zone and were exposed to violent treatment by the Greek Border Guard. More than 5,000 refugees stayed in the border zone between February 27 and March 27, 2020. Following the outbreak of the COVID-19 pandemic, Turkish guards opened the Turkish borders and reaccepted the refugees while declaring that when the pandemic is over, Turkey will open the European borders. For a more detailed analysis, see Association for Migration Research Video Series, "What Happened at the Turkish-Greek Border," available at https://www.youtube.com/watch?v=BAso-OXzsoY&t=78s. Last accessed on July 13, 2020.

References

Akdoğan, Muzaffer. 2018. "Avrupa Birliği'nin Sınırlarına Dayanan Mülteci Krizi ve Yönetici." *Uluslararası İlişkiler ve Diplomasi Dergisi* 1 (1): 48–74.

Apostolova, Raia. 2015. "Economic vs. Political: Violent Abstractions in Europe's Refugee Crisis." FocaalBlog. https://www.focaalblog.com/2015/12/10/raia-apostolova-economic-vs-political-violent-abstractions-in-europes-refugee-crisis/#more-1692.

Ataç, Ilker, Gerda Heck, Sabine Hess, Zeynep Kaşlı, Philipp Ratfisch, Cavidan Soykan, and Bediz Yılmaz. 2017. "Contested B/Orders. Turkey's Changing Migration Regime." *Movements* 3 (2).

Berg, Mette Louise, and Elena Fiddian-Qasmiyeh. 2018. "Introduction to the Issue: Encountering Hospitality and Hostility." *Migration and Society* 1.

Biehl, Kristen Sarah. 2015. "Governing through Uncertainty: Experiences of Being a Refugee in Turkey as a Country for Temporary Asylum." *Social Analysis* 59(1): 57–75.

Biner, Özge. 2014. "From Transit Country to Host Country: A Study of Transit Refugee Experience in a Border Satellite City, Van, Eastern Turkey." In *Migration to and from Turkey: Changing Patterns and Shifting Policies*, edited by Ayşem Biriz, Tokay Karaçay, and Ayşen Üstübici. Istanbul: The Isis Press.

Bojadžijev, Manuela, and Sandro Mezzadra. 2015. "'Refugee Crisis' or Crisis of European Migration Policies?" FocaalBlog. http://www.focaalblog.com/2015/11/12/manuela-bojadzijev-and-sandro-mezzadra-refugee-crisis-or-crisis-of-european-migration-policies/.

Cantat, Céline. 2016. "Rethinking Mobilities: Solidarity and Migrant Struggles Beyond Narratives of Crisis." *Intersections: East European Journal of Society and Politics* 2 (4): 11–32.

———. 2020. "Governing Migrants and Refugees in Hungary: Politics of Spectacle, Negligence and Solidarity in a Securitising State." In *Politics of (Dis)Integration*, edited by Sophie Hinger and Reinhard Schweitzer, 183–200. Springer Open.

Cavdar, Ayşe. 2016. Oktay Durukan'la Söyleşi: Mülteci kimdir, bize ne öğretir? [Interview with Oktay Durukan: Who is a Refugee, What Does She Teach Us?]. *Saha Helsinki Yurttaşlar Derneği Süreli Yayını*, 20–26.

Çavuşoğlu, Mevlüt. 2016. "Turkey and the Syrian Refugee Crisis: An Example for Humanity." *Turkish Policy Quarterly* 15 (3): 17–24.

Chatty, Dawn. 2010. *Displacement and Dispossession in the Modern Middle East.* Cambridge: Cambridge University Press.

Czajka, Agnes. 2015. "Migration in the Age of the Nation-State: Migrants, Refugees, and the National Order of Things." *Alternatives: Global, Local, Political* 39 (3): 151–63.

De Genova, Nicholas. 2013. "Spectacles of Migrant 'Illegality': The Scene of Exclusion, the Obscene of Inclusion." *Ethnic and Racial Studies* 36 (7): 1180–98.

———. 2018. "The 'Migrant Crisis' as Racial Crisis: Do Black Lives Matter in Europe?" *Ethnic and Racial Studies* 41 (10): 1765–82.

Dündar, Fuat. 2015. *İttihat ve Terakki'nin Müslümanları İskân Politikası (1913–1918).* Istanbul: İletişim Yayınları.

Erdoğan, Murat. 2016. "Avrupa'da Yaşanan 'Mülteci' Krizi Değil 'İnsanlık' Krizi." *Anadolu Ajansı*, October 7, 2016. https://www.aa.com.tr/tr/analiz-haber/avrupada-yasanan-multeci-krizi-degil-insanlik-krizi/660204.

Fassin, Didier. 2011. *Humanitarian Reason: A Moral History of the Present.* 1st ed. Berkeley: California: University of California Press.

Genç, Fırat, Gerda Heck, and Sabine Hess. 2018. "The Multilayered Migration Regime in Turkey: Contested Regionalization, Deceleration and Legal Precarization." *Journal of Borderlands Studies* 34 (4): 489–508.

Heck, Gerda, and Sabine Hess. 2017. "Tracing the Effects of the EU-Turkey Deal: The Momentum of the Multi-Layered Turkish Border Regime." *Movements* 3 (2): 35–56.

Hess, Sabine, and Bernd Kasparek. 2017. "De- and Restabilising Schengen: The European Border Regime after the Summer of Migration." *Cuadernos Europeos de Deusto* 1 (56): 47–77.

İçduygu, Ahmet. 2011. "The Irregular Migration Corridor between the EU and Turkey: Is It Possible to Block It with a Readmission Agreement?" *Technical Report, EU-US Immigration Systems*, 2011/14. European University Institute Research Repository. https://cadmus.eui.eu/handle/1814/17844.

İktisadi Kalkınma Vakfı. 2016. "Mülteci Krizi Ekseninde Türkiye-AB İşbirliği." İktisadi Kalkınma Vakfı Yayınları, Yayın No. 289.

Kallius, Annastiina, Daniel Monterescu, and Prem Kumar Rajaram. 2016. "Immobilizing Mobility: Border Ethnography, Illiberal Democracy, and the Politics of the 'Refugee Crisis' in Hungary." *American Ethnologist* 43 (1): 25–37.

Karadağ, Sibel. 2019. "Extraterritoriality of European Borders to Turkey: An Implementation Perspective of Counteractive Strategies." *Comparative Migration Studies* 7 (1): 1–16.

Kilberg, Rebecca. 2014. "Turkey's Evolving Migration Identity." *Migration Policy Institute.* Available at https://www.migrationpolicy.org/article/turkeys-evolving-migration-identity.

Li, Tania Murray. 2010. "To Make Live or Let Die? Rural Dispossession and the Protection of Surplus Populations." *Antipode* 41: 66–93.

Malkki, Liisa H. 1996. "Speechless Emissaries: Refugees, Humanitarianism, and Dehistoricization." *Cultural Anthropology* 11 (3): 377–404.

Marfleet, Philip. 2016. "States of Exclusion." *Socialist Worker: News and Analysis.* Available at https://socialistworker.co.uk/socialist-review-archive/states-exclusion/.

Netherlands Helsinki Committee. 2019. "Defending Human Rights in Turkey: Cavidan Soykan." https://www.nhc.nl/cavidan-soykan/.

New Keywords Collective. 2016. "Europe/Crisis: New Keywords of 'the Crisis' in and of 'Europe.'" Near Futures Online 1 "Europe at a Crossroads." http://nearfuturesonline.org/wp-content/uploads/2016/01/New-Keywords-Collective_12.pdf.

Nyers, Peter. 1998. "Refugees, Humanitarian Emergencies, and the Politicization of Life." *Refuge Canada's Journal of Refugees/Revue Canadienne sur les Refugiés* 17 (6).

Özçürümez, Saime, and Nazli Şenses. 2011. "Europeanization and Turkey: Studying Irregular Migration Policy." *Journal of Balkan and Near Eastern Studies* 13 (2): 233–48.

Özden, Senay. 2013. *Syrian Refugees in Turkey*. Migration Policy Centre, MPC Research Report, 2013/05. https://hdl.handle.net/1814/29455.

Öztürk, Aysel. 2017. "The Production of Refugee Subjectivities in the State Discourse: The Case of Syrian Refugees in Turkey." Unpublished master's thesis, İstanbul Şehir University.

Rajaram, Prem Kumar. 2015. "Beyond Crisis: Rethinking the Population Movements at Europe's Border." FocaalBlog. http://www.focaalblog.com/2015/10/19/prem-kumar-rajaram-beyond-crisis/#more-1491.

Roy, Ananya, and Nicholas De Genova. 2018. "New Poverty Politics for Changing Times: What Emerging Nationalist Populisms Mean for Poverty and Inequality." Interview transcript. Relational Poverty Network. January 31. https://www.dropbox.com/s/39w7488yvcc9est/Roy%20and%20De%20Genova_RPN%20New%20Poverty%20Politics%20Interview_Final.pdf?dl=0.

Rozakou, Katerina. 2012. "The Biopolitics of Hospitality in Greece: Humanitarianism and the Management of Refugees." *American Ethnologist* 39 (3): 562–77.
Rygiel, Kim, Feyzi Baban, and Suzan Ilcan. 2016. "The Syrian Refugee Crisis: The EU-Turkey 'Deal' and Temporary Protection." *Global Social Policy* 16 (3): 315–20.
Sarı, Elif, and Cemile Gizem Dinçer. 2017. "Toward a New Asylum Regime in Turkey?" *Movements* 3 (2): 59–79.
Şen, Yusuf Furkan, and Gözde Özkorul. 2016. "Türkiye – Avrupa Birliği İlişkilerinde Yeni Bir Eşik: Sığınmacı Krizi Bağlamında Bir Değerlendirme [A New Threshold in Turkey - European Union Relationships: A Review Within the Context of Refugee Crisis]." *Göç Araştırmaları Dergisi* 2 (2): 86–119.
Soykan, Cavidan. 2010. "The Migration-Asylum Nexus in Turkey." *Enquire* 3 (1): 1–18.
Ticktin, Miriam I. 2011. *Casualties of Care: Immigration and the Politics of Humanitarianism in France.* Berkeley: University of California Press.
UNHCR. 2019. "Turkey: Key Facts and Figures." November 2019. https://www.unhcr.org/tr/wp-content/uploads/sites/14/2020/01/73154.pdf.
Watenpaugh, Keith David. 2015. *Bread from Stones: The Middle East and the Making of Modern Humanitarianism.* Berkeley: University of California Press.

6

The Transformation of Citizens to Refugees: The Case of the Rohingya of Myanmar

Tasneem Siddiqui and C R Abrar

In 2013, the United Nations characterized the Rohingya population as one of the most persecuted minorities in the world (OHCHR 2017). Since 1978, after thirty years of independence, the military elite of Myanmar has increasingly claimed that the Rohingya do not belong in Myanmar. Since then, the military elite has fostered a hostile environment that has forced many Rohingya people to flee their ancestral home in different phases. By August 2017, Myanmar was able to expel 90 percent of the Rohingya population. Most took refuge in Bangladesh, while others managed to migrate to Saudi Arabia, Malaysia, Indonesia, India, and a few other countries. This paper aims to understand the processes that led to turning many Rohingya people into refugees.

The chapter explores the role of history in shaping Rohingya refugee movements. It attempts to locate the implications of colonial rule and the decolonization process. It also aims to understand the effect of postcolonial political developments and nation-building exercises on different ethnoreligious groups in Myanmar. The chapter examines the role of external actors as well.

The chapter is divided into seven sections. The first presents the main objectives. It also deals with conceptual issues and methodology. The second presents the historical backdrop that shaped migration flows between Myanmar and Bangladesh. The third focuses on the impact of colonial rule and the process of decolonization. The fourth scrutinizes the nation-building exercises of successive regimes that contributed to disenfranchising the

Rohingya population and transforming them into refugees. The fifth gives an account of refugee flows from Myanmar to Bangladesh from 1978 until the present. The sixth looks into the role of global politics in creating or exacerbating the protracted refugee situation. Finally, the seventh section draws some conclusions.

Conceptual Issues

Neither Bangladesh nor Myanmar has specific laws pertaining to refugees. While discussing the Rohingya plight, this paper follows the UN definition of refugees as it is an internationally acknowledged standard. The UN 1951 Convention defined a refugee as

> any person who [. . .] as a result of events occurring before 1 January 1951 and owing to well-founded fear of being persecuted for reasons of race, religion, nationality, membership of a particular social group or political opinion, is outside the country of his nationality and is unable or, owing to such fear, is unwilling to avail himself of the protection of that country; or who, not having a nationality and being outside the country of his former habitual residence as a result of such events, is unable or, owing to such fear, is unwilling to return to it (Article 1(A)(2) of 1951 Refugee Convention).

The research draws on the historical-structuralist framework in explaining the gradual transformation of the Rohingya population of Myanmar into refugees. The historical-structuralist approach is not a single theory but contains a diverse range of ideas. Nonetheless, the historical structuralists offer an understanding of the wider backdrop, including geopolitical as well as socioeconomic structures within which individuals or countries operate and their role in shaping different outcomes. Singer (1991) and Stern (1988) have explained the basic ideas of the structuralist school of thought with regard to migration.[1] Their basic premise is that broader social and historical forces create conditions of widespread inequalities in accessing resources, exercising political power, and attaining prestige within a community and across communities, which both stimulate and constrain human movement. International migration, including refugee flows in general, is, therefore, a natural consequence of capitalist market formation in the developing world. The creation of colonies was the most widely practiced system that allowed colonial powers to pursue their economic interests in different parts of the world. The economic and strategic interests of the former colonial powers are also pursued, even in postcolonial situations, by creating new relations and interactions with their former colonies.

In recent times, while understanding refugee flows or voluntary labor migration, most researchers have concentrated on social, economic, and, to an extent, political factors in the areas of origin. These studies neglect the broader historical legacy: the impact of colonial rule and the interplay of economic, strategic, and political interests of regional and global powers in creating and sustaining forced or voluntary migration.

In this chapter, we want to pursue the hypothesis that turning the Rohingya into refugees and prolonging their status as such is linked to the history of state formation, colonialism, and capture of state power by the military representing the dominant ethnic group and the self-interest of the global powers. The organic process of state formation in both Myanmar and Arakan was disrupted by colonization. For administrative purposes, the British treated Burma and Arakan as a single unit, whereas for political expediency, it pursued a policy of "divide and rule." Seeds of conflict among various ethnoreligious groups across Burma, including Arakan, were already sown by the time of the British exit. In the postcolonial period, while political leadership in some countries attained a reasonable degree of success in diffusing interethnic tension and conflicts by accommodating plurality and diversity, other countries failed to do so. In most of the latter cases, the dominant subnational, ethnic, or religious group pursued its own interests to the detriment of other groups.

In newly independent Burma, Rohingya and Rakhine Buddhists of Arakan were initially treated as inseparable from the fabric of Myanmar, participating in the nation-building process. The military takeover of Burma in 1962 and the regime's espousal of faith- and ethnicity-based nationalism excluded many ethnic groups, including the Rohingya of Arakan. They were labeled as the "other." Besides, the military pursued a policy of isolation for six decades, from 1962 to 2011. When this resource-rich and strategically significant country gradually came out of isolation and became integrated into the global economy, regional and global powers pursued their self-interests in forging their relationship with Myanmar.

Precolonial Past

The Rohingya people form a distinct ethnoreligious community that has been living in the Arakan State of Myanmar for centuries. Most of them are Muslims, with only a fraction belonging to the Hindu faith. The people of Arakan were exposed to Islam through Arab merchants in the first millennium (Malik 2000, 22). Many inhabitants of the region subsequently converted to Islam.

In AD 951, an Arakanese king occupied the Chattogram region, and the name of the region was eventually derived from his name (Malik 2000, 22). Subsequently, the rulers of Bengal succeeded in re-establishing their control over Chattogram. Muslim contacts with Rakhine were firmly established when the Mughal government of Bengal aided the deposed king Narameikhla to regain his throne in 1430. In 1785, the Burman king annexed Arakan, and the Muslims of Arakan started experiencing discriminatory treatment (Chan 2005, 399). During this time as well, some Arakan Muslims and Rakhine Buddhists took refuge in Chattogram. The absorption of Arakan brought Burma into direct contact with the British colonial rulers of India. The raids of Arakanese refugees based in Chattogram across the border led to Burman army incursions into British Indian territory, resulting in the first Anglo–Burmese War (1824–26). This led to the annexation of Arakan and Tenasserim by the British. The subsequent second (1853) and the third (1855) Anglo–Burmese Wars resulted in British colonial control over the entirety of Burma (Abrar 1996, 5). The British ruled Burma as a separate colony known as British Burma, with Arakan as one of the administrative divisions. Later, when the Second World War began, the Burmans sided with the Japanese Army. The Arakanese communities were divided in their loyalties; the Rohingya were loyal to the British, and their Buddhist Rakhine counterparts sided with the Japanese. After the Second World War, the British decolonization process gathered pace. Though Britain promised the Rohingya a separate land—a Muslim National Area—in exchange for their support during the war, it gave independence to Burma without securing any such area within that state (Yegar 1972, 96). The Rohingya felt betrayed.

The State and Nation-Building Process in Burma

After independence, as the Burmese state began to establish its control over Arakan and adjoining regions, the Buddhist Arakanese organized a resistance movement. In order to win over the Rohingya support, the Burmese authorities granted them special status (Zarni and Brinham 2019). The founding nationalist leaders of Burma, including General Aung San, framed the constitution to create a multiethnic, multicultural, and multireligious identity of Burma. The *Encyclopedia of Burma* (1964, 90) clearly states that "of the 400,000 to 500,000 inhabitants of Mayu Border District, 75% are Rohingya ethnic people who are Muslims. The rest (are) made up of Rakhine, Dai Nat, Myo and Kamee" ethnic groups.

During the semi-liberal rule of the Anti-Fascist People's Freedom League, the emphasis was to create national integration through ethnic union. As the

Rohingya stood up against the separatist Muslims of Bengal, they were officially praised as a national group in the mid-1950s by President U Nu (Poling 2014).

Social scientists of the Global North were extremely enthusiastic about the political, social, and economic development potentials of the newly independent countries. They prescribed models for these countries to achieve such goals. "Modernization" became the single most important term in planning future economic growth and political stability of the newly independent countries. These studies identified the armed forces of these states as one of the agents of modernization (Finer 1962; Huntington 1968). By the early 1960s, many newly independent countries began experiencing military rule. In 1962, Burma also experienced a military takeover. The nationalist political elite were marginalized. In the ensuing years, the military regime dismantled the national identity from multicultural, multiethnic, and multifaith identity to a narrower one that established the dominance of the single ethnic identity of the Bamars (Zarni and Cowley 2014, 82). Along with Bamars, Shans, Kachins, and Karens, a few other ethnic groups were recognized. These perceived agents of modernization, however, for the first time, excluded some groups from the nation-building process. Politics of "othering" began. The Rohingya people were one of the ethnic groups who were deemed as "others" or "outsiders."

The Process of "Othering"

The military government introduced the idea of *taingyintha*[2] that recognized 135 national races. Nonetheless, for the military rulers, Bamar was the dominant ethnicity, and the other 134 races were classified as minorities who should accept Bamar dominance. The overriding mindset was to cleanse the nation through "systematic attempts to subjugate some ethnic minorities whilst removing others (such as Rohingyas) from the national fold" (Zarni, Brinham, and Cowley 2017). Any group that was not considered indigenous and, therefore, foreign was perceived as a threat to national solidarity. The Rohingya were excluded from the list and were identified as Sunni Muslims of Bengali ethnicity, not part of Burma. The Burmese armed forces continued to uphold colonial-era resentment and fear of people of Indo-Aryan[3] origin, and more specifically, Muslims.

The project of "othering" the Rohingya took a violent turn in 1978, when Ne Win, former president of then Burma (Myanmar today), launched a centrally organized operation in both the southern and northern Rakhine states. Known as Operation Dragon King, the joint forces of immigration, religious affairs, police, courts, army, navy, and intelligence agencies were empowered to conduct surprise immigration checks.

Refugee Flows

In independent Myanmar, the first major refugee flow began in late 1978 after Operation Dragon King. The operation was so violent that in 1978, a total of 277,938 Rohingya people fled between February 12 and June 3 into the neighboring Bangladesh (Nyunt 2016, 21–43). Following bilateral negotiations between Bangladesh and Myanmar, the Rohingya refugees returned within a few months.

Burma enacted its citizenship law in 1982, a law that was used as an instrument to deny the Rohingya and their future generations their right to Burmese citizenship (Lewa 2009, 11). This act recognizes three categories of citizens: full citizen, associate citizen, and naturalized citizen. The first group of citizens comprises those with Buddhist roots, who are issued pink cards. The second group refers to those who can produce evidence of entry before 1948 and speak at least one Burmese language and/or were born in the country before 1948. They are labeled as associate citizens and issued blue cards. The third group is naturalized citizens entitled to green cards. The Rohingya were categorized as temporary residents and received a Temporary Registration Card (white card). The white card does not state their place of birth, making them ineligible for citizenship. More importantly, the 1982 Citizenship Act does not recognize the Rohingya as one of the national ethnic groups of the country.

Along with the framing of the new citizenship laws, the military authorities also began instituting discriminatory laws and administrative orders against the Rohingya. These laws and orders helped the Burmese state forcibly confiscate and evict them from their ancestral land. This discriminatory and arbitrary treatment resulted in forced labor conditions for military projects, forced relocations, conditions to secure permissions for marriage, restrictions on movements and practice of faith, etc. Their farms were periodically ransacked; men were tortured, women were raped, and this resulted in early marriage among Rohingya children becoming common, as well as repeated pregnancies to protect themselves from rape.

The successions of persecution and often violence meted out to them resulted in a search for refuge again in Bangladesh in 1991–92. This time, 256,193 Rohingya people crossed the border to Bangladesh (Zarni and Brinham 2019, 53). Following the 1992 agreement between the two governments, some 230,000 refugees were repatriated to Myanmar with the assistance of UNHCR.[4] A small number of refugees (21,600) stayed on (MSF 2002, 23).

As the conditions in Myanmar did not change and the persecution of the Rohingya continued, by July 1996, a fresh influx of them arriving in

Bangladesh was reported. The government did not accept them as refugees and refused to give them shelter in UNHCR-managed refugee camps. Due to religious affinity, a few Gulf and Southeast Asian countries were sympathetic to their dilemma, such as Saudi Arabia and Malaysia, which became the two major destinations they sought out.

In May 2012, Bangladesh experienced another large inflow of Rohingya people. This flow was the result of indifference by the Myanmar authorities toward the violence committed by the Buddhist majority, which resulted in the killing of at least fifty Rohingya people and the displacement of 30,000. The event was triggered following the alleged rape and murder of a Rakhine woman by three Rohingya men. Bangladesh refused to accord shelter to the persecuted Rohingya. By then, the Indian government had labeled them as Muslim terrorists, supported by the Inter-Services Intelligence wing of Pakistan. Some Bangladeshi intellectuals also subscribed to this position and viewed the Rohingya refugees as a security threat to Bangladesh. During this period, many Rohingya people found shelter by irregular means using their preexisting social networks. Their number is also significant.

In 2013, the government of Bangladesh carried out a census with the help of the Bangladesh Bureau of Statistics. The figure is not officially available. However, 200,000–300,000 Rohingya people may have entered the country informally through marriage and other methods, trying to secure their stay in Bangladesh. Some explored avenues for migration to a third country. One of the paths used to flee to third countries was securing Bangladeshi passports (Siddiqui 2018, 130).

It was during the communal violence against the Rohingya in 2012 that the voices of the diaspora began to emerge to a great extent. Many organizations concerned with the Rohingya diaspora engaged in a global campaign against the atrocities committed by the Myanmar security forces, demanding the restoration of full citizenship rights and an end to all discriminatory laws against members of the community. The Rohingya diaspora mobilized students on campuses around the world and utilized social media to gain attention and support.

After a prolonged struggle for democracy, Aung San Suu Kyi was voted into power in 2015. However, there has not been any discernible change in government policy toward the Rohingya. Instead, violence against them kept breaking out, and more of them were pushed to leave for Bangladesh. The expulsion of the Rohingya following massive atrocities in August 2017 resulted in a record refugee influx to Bangladesh. The International Criminal Court has observed that the Myanmar government's actions against the

Rohingya population had borne "the hallmarks of genocide" (R.C. 2018), while others termed it an "ethnic cleansing" (UN News 2018; Human Rights Watch 2013). In its final session held in Kuala Lumpur in September 2017, the People's Tribunal of State Crimes of Myanmar unequivocally termed it a "genocide" (Permanent Peoples' Tribunal 2018, 48).

With 1.1 million Rohingya inhabitants, Bangladesh hosts the largest number of Rohingya people, but other countries are also hosting them. Apart from those in Bangladesh, 470,000 Rohingya people live in Saudi Arabia, 450,000 in Pakistan, 200,000 in Malaysia, 50,000 each in India and the UAE, and 5,000 each in Thailand, Australia, and the USA. In addition, the EU, Jordan, Indonesia, Canada, Nepal, Sri Lanka, and Japan host more than 8,500 Rohingya people (Sorwar 2019). Those who came to Bangladesh after August 25, 2017, were registered with the relevant government authorities, which entitled them to secure shelter, food, and medical services.

In October 2017, the governments of Bangladesh and Myanmar signed two agreements: one on security and the other on border cooperation. These two governments have also agreed that they would take concrete measures for the safe, honorable, and secure return of the Rohingya. Unfortunately, under the agreed-upon terms, the final verification rested with Myanmar. The agreement also could not ensure adequate international monitoring. Another important gap in the agreement is the absence of any timeframe to complete the repatriation process. As per the agreement, repatriation was to begin as early as January 23, 2018. However, no repatriation has taken place thus far.

By 2017, the Rohingya diaspora became more organized and started campaigning to expose how northern Arakan was being cleared of the Rohingya through a sustained process of ethnic cleansing. Diaspora organizations were instrumental in establishing the International Peoples' Tribunal on Myanmar's State Crimes against Rohingya and Other Ethnic Minorities in September 2017. The tribunal ruled that Myanmar was guilty of genocide against the Rohingya and the Kachins. These groups, including Burmese dissident scholars, formed the Free Rohingya Coalition (FRC) that called for boycotts, economic and diplomatic sanctions, and other acts of solidarity. This group is engaged in an international campaign through conducting evidence-based research, hosting conferences, publishing opinion pieces, appearing in electronic media, and rallying the support of eminent intellectuals.[5] It has been persistent in raising the demand that the UN Security Council take effective action to stop the genocide, ensure justice, and hold perpetrators accountable. In May 2020, the Burmese Rohingya Organization

UK (BROUK) mobilized a court in Argentina to pursue a case against Myanmar's leader, Aung San Suu Kyi, and senior officers in the military over the genocide and persecution of the Rohingya community (Kamruzzaman 2020). Some Rohingya people in diaspora articulated the demand for the "protected return to protected homeland" (FRC, n.d.).

Geopolitics

Myanmar and Bangladesh are two neighboring countries sharing a 270-kilometer border. The border starts in the north at the tripoint with the Indian state of Mizoram on land before reaching the Naf River and subsequently terminating southward in the Bay of Bengal. On the Bangladeshi side of the border, there are the Chittagong Hill Tracts and Cox's Bazar District. On the Myanmar side, there is Arakan, now known as Rakhine State.

Both Bangladesh and Myanmar were part of the British colonial empire. Myanmar became an independent state in 1948, whereas Bangladesh gained independence twice: first from Pakistan, then in 1971, after a nine-month war of independence that severed its ties with Pakistan, the new state of Bangladesh was born.

Both countries have geopolitical significance in the region. With its natural resources, Myanmar is one of the most energy-rich countries in the region. It is the gateway to Southeast Asia for South Asian countries. Myanmar has a long-standing relationship with China, with China being the only country that had access to its economy and made huge investments there during the period when it isolated itself from the world. Given their shared borders, maintaining good relations with Myanmar is significant for India, especially for the sake of their mutual investments and trade. These relations are also important to counterbalance China's influence on the country. In addition, European countries suddenly became interested in benefiting from trade and investment opportunities when six decades of Myanmar's isolation ended. Balancing China has also been an important part of European strategic interests.

Bangladesh has one of the fastest-growing economies in the world and has the potential to grow further by developing its "blue economy"[6] by harnessing resources from the Bay of Bengal. In recent years, India has expressed interest in investing in transportation infrastructure by developing its land and rail routes. In order to facilitate the movement of goods to its northeastern states, India already has transshipment and other arrangements with Bangladesh. Bangladesh is also gaining importance among the countries of the Global North, as the Bay of Bengal intersects with the route for the export of gas and other resources. China has also invested heavily in Bangladesh in different sectors. Therefore, Bangladesh has been performing a difficult balancing act vis-à-vis its relations with China and India. Bangladesh and Myanmar could benefit from mutual gains if friendly and neighborly relations can be maintained.[7]

Global Actors: National Interest versus Humanitarianism

The global actors of the North and the great powers of Asia have paid little attention to Myanmar's ongoing discrimination, disenfranchisement, persecution, and, finally, genocide against the Rohingya population. Perhaps this lack of engagement is guided by their pursuit of strategic security, trade, investment, and political interests in Myanmar. The pursuit of these national interests by the Global North and the great powers of Asia with no regard for the rights of national minority groups has aided Myanmar's continued persecution and expulsion of Rohingya people and other minority groups. The UN General Assembly and the UN Council for Human Rights have routinely conveyed grave concerns about the ongoing violations of Rohingya rights in Myanmar. In a report to the Human Rights Council in 2019, the Independent International Fact-Finding Mission on Myanmar concluded "on reasonable grounds that the Government's acts continue to be part of a widespread and systematic attack against the remaining Rohingya in Rakhine State, amounting to the crimes against humanity of inhumane acts and persecution" (Human Rights Council 2019a, 7; Human Rights Council 2019b).

The UN Security Council members, nonetheless, have largely remained passive. Their inaction was noted in Gert Rosenthal's report to the UN Secretary-General. He opined that "the options of the UN to address the challenge in a manner consistent with its values and principles are often rather limited" (Rosenthal 2019, 11). He further observed that Article 99 of the UN Charter authorizes the Secretary-General to "bring to the attention of the Security Council any matter that in his opinion may threaten the maintenance of international peace and security" (United Nations, n.d.).

Countries of the Global North have shown concern for the Rohingya plight and have extended humanitarian support, though they hardly act to expedite durable solutions. In October 2017, the United States withdrew military aid to some Myanmar units responsible for the displacement and imposed travel restrictions on some current and former senior leadership of the Myanmar military. However, calls for sanctions by some congressional members and human rights activists were not heeded on the grounds that they would be an ineffective method to bring about their desired objectives and might also jeopardize US influence in Myanmar. In December 2017, the US blacklisted the chief of the Myanmar army's Western Command, Major General Maung Maung Soe, freezing his US assets and barring all Americans from engaging in business transactions with him. In August 2018, the United States Treasury Department enacted economic sanctions on four Myanmar military and police commanders for their role in "ethnic cleansing" (US Department of Treasury 2018). On December 10, 2019, the US imposed sanctions on four Myanmar military leaders, including the commander-in-chief, for alleged human rights abuses against the Rohingya and other minorities (Lewis and Psaledakis 2019). The Gambian case at the International Court of Justice against Myanmar, accusing it of genocide, led the US to impose sanctions against Myanmar army chief Min Aung Hlaing (Lewis and Psaledakis 2019). The US has provided nearly $700 million for programs to support the Rohingya refugees in Bangladesh.

China has a major influence over Myanmar and has been an important security partner of it. Myanmar's control over the Rakhine region is also in the strategic and economic best interests of China, particularly for energy security (Yunnan Daily 2018). A proposed oil pipeline will connect Myanmar's Kyaukpyu port with southwestern China. This facility will enable China to import gas and oil from the Middle East, significantly shortening the shipping time compared to the current route through the Malacca Straits. China defends its policy with the rhetoric of noninterference in the internal affairs of another country.

Strategic, political, and economic considerations also figure prominently in the Indian response to the Rohingya crisis. A fundamental consideration in the Indian approach is its geopolitical interest in countering protracted Chinese influence in Myanmar. Myanmar is the only Southeast Asian country that shares land and maritime borders with India. India perceives Myanmar as a country that can play a pivotal role in establishing the ASEAN–India Free Trade Area. It is in this context that India invested in the India-Myanmar-Thailand Trilateral Highway and the Kaladan Multimodal Transit Transport project. As a response to the Chinese initiative to build a seaport in Kyaukpyu, India invested in constructing a seaport in Sittwe. Military cooperation in the form of training, joint military and naval exercises, Indian export of military hardware, including a refurbished submarine to Myanmar, and recent high-level visits by senior generals are indications of a major boost in Indo-Myanmar multifaceted cooperation.

The Islamophobic nature of the current Indian government also contributes to its Rohingya policy. Even the interim order of the International Court of Justice made little difference. As of 2021, India had provided $25 million in aid to Myanmar to build prefabricated houses, schools, and a medical facility in the temporary camp being built for returnee Rohingya people on the Myanmar side. Over the last few years, India has deported Rohingya people to Myanmar (Hussain 2019). India has also been engaged in pushing back Rohingya people to Bangladesh without Dhaka's permission. From May 2018 to January 2019, over 1,300 Rohingya people were sent to Bangladesh (Bhuiyan 2019). The EU imposed an arms embargo on equipment that may be used by the Myanmar police or military or for internal repression or monitoring communications, as well as a prohibition on the provision of military apparatus that could be used in security crackdowns in Myanmar. The sanctions also imposed travel bans and asset freezes on fourteen top military and border officials "for serious human rights violations, or association with such violations" against Myanmar's Muslim Rohingya minority. The EU also refuses to cooperate with or provide training to the Myanmar military (European Council 2019).

Along with the above measures to exert pressure on Myanmar, in March 2020, the EU announced an allocation of €22 million to support the most vulnerable groups among refugees and host communities in Bangladesh (European Commission 2020). Another €3.5 million has been allocated to build the local communities' capacities to prepare for and manage natural disasters. Since 2017, the European Union has provided over €150 million in humanitarian aid to respond to the Rohingya crisis both in Myanmar and Bangladesh. It covers all types of refugees: those who have been staying in Bangladesh for many years and newly arrived registered refugees. The program also includes members of host communities living close to the refugee settlements.

As a regional body, the role of the Association of Southeast Asian Nations (ASEAN) has been inadequate in addressing the Rohingya problem. Though member states are committed to "enhance good governance and the rule of law and to promote and protect human rights and fundamental freedoms," as per Article 1.7, and respond effectively "to all forms of threat, transnational crimes and trans-boundary challenges" (ASEAN n.d., 3), thus far the regional body has refrained from taking action against Myanmar's atrocities in the Rakhine region. Its engagement on the Rohingya issue has been largely to provide humanitarian assistance and support repatriation efforts.

This lack of action stems from the regional grouping's uncritical endorsement of the Myanmar government's narrative that it is engaged in suppressing an ethnic insurgency in Rakhine and, thus, an internal security matter. So far, the ASEAN Chairman's Statements and Foreign Ministers' Joint Communiques have made no reference to the Rohingya situation but have extended support to the Myanmar government's own Commission of Enquiry to investigate human rights violations in Rakhine. Choudhury (2018) observes that such an approach by the ASEAN would result in the "growth of new transnational trafficking and militant networks."

The Organization of Islamic Cooperation (OIC) is uniquely positioned to advocate the Rohingya cause. In the past, the OIC stood up for the rights of Palestinian and Kashmiri peoples. As early as January 2017, the organization criticized Myanmar's policy toward the Rohingya and demanded an end to state violence and access to the Rakhine region (OIC 2017). Along with other countries such as Turkey and Malaysia, the OIC was forthcoming in providing material, political, and humanitarian assistance to Bangladesh. It appointed a former Malaysian foreign minister as its envoy on Rohingya issues to exert moral pressure on the international community. The OIC Council of Ministers also established the Ad Hoc Ministerial Committee on Accountability and Human Rights Violations against the Rohingya. It was with OIC's explicit support that Gambia made a submission to the International Court of Justice invoking the Genocide Convention (Van de Berg 2019). However, the policies of some OIC members are frequently contradictory.

Among OIC countries, Saudi Arabia has been notably sympathetic to Rohingya genocide survivors for decades. More than 50,000 Rohingya people have been residing in Saudi Arabia for generations. This has changed since mid-2020, when Saudi Arabia abdicated its responsibility by attempting to deport them to a different country, in this case, Bangladesh. Saudi Arabia's request was tied to future access of Bangladeshi workers to the Saudi labor market (Islam 2020).

There is a major difference between the state actors and non-state actors of the Global North. Non-state actors such as civil society organizations, particularly the students at different universities in the Global North, have extended moral support to the persecuted Rohingya. By holding rallies, exhibitions, and seminars, they have been articulating the demand for durable solutions to the Rohingya refugee situation. In many instances, the awards and accolades bestowed on Burmese State Counsellor Suu Kyi for her role in restoring democracy in Myanmar have been rescinded (Reuters 2018; BBC 2018).

Conclusion

This article concludes that turning Rohingya people from bona fide citizens of Myanmar into refugees cannot be explained by a single discipline-based analysis. It cannot be understood by analyzing Myanmar's political history since the military takeover of power in 1962 either. Espousing a historical-structuralist approach allowed this research to develop a broad and holistic interdisciplinary framework that combines elements of history, the role and consequences of events at different historical junctures, and the contemporary politics of Myanmar and its relations with other international actors. Such an approach considered precolonial state formation processes, the perfunctory integration of preexisting diverse political entities effected through colonial rule, and the colonial administration's pursuit of "dividing and ruling" ethnoreligious groups. It let us examine the impact of the Second World War on different communities and arbitrary decolonization processes in breach of wartime promises to establish an autonomous Muslim National Area. It could rationally place the military takeover of the Myanmar state and subsequent dominance of one ethnic group, the Bamar in Myanmar, in the context of analyzing successive moves of discriminatory laws and provisions targeting the Rohingya and other minority groups. All this ultimately led to the genocide committed against the Rohingya people and their eventual exodus.

The Global North, the large Asian powers, and regional organizations played their part in creating and sustaining a system that disenfranchised the Rohingya. They did so by prioritizing their own strategic trade and investment interests. As the country most adversely affected by the Rohingya influx, Bangladesh has its own responsibility to shoulder. For a long time, Bangladesh tried to address the problem through bilateral negotiations. Until 2018, it did not internationalize their plight. All these factors contributed to the expulsion of the Rohingya from their ancestral land.

Civil society organizations, human rights advocates, student activists, and Rohingya people in diaspora all over the world have begun to articulate the demand for justice, accountability, and reparation. The solution they demand is the "protected return" of the Rohingyas "to protected homeland" (Abrar 2018). In 2018, the UN General Assembly and the government of Bangladesh also proposed the establishment of a safe zone for the Rohingya in Rakhine State (Kamruzzaman 2019). It is obvious that a long battle is still ahead before a protected homeland for the Rohingya can be established.

Notes

1 For further discussion of this, see Massey et al. (1998) and Bilsborrow, Oberai, and Standing (1984).

2 *Taingyintha* means "ethnic groups" or "indigenous races" (see Karen Human Rights Group 2020).

3 People who migrated from central to south Asia in the second millennium BC.

4 In 1989, the country was renamed "Myanmar" by the government. In the following section, we use the name Myanmar.

5 This section is based on information gathered in a virtual consultation with members of the FRC, which took place on December 22, 2020. It was attended by Hafsar Tameesuddin, Rohingya rights activist in New Zealand; Ali Johar, filmmaker in India; Sharifa Shakira of the Rohingya Women Development Network in the US; Maung Zarni, Burmese dissident and Rohingya genocide scholar based in the UK; Nay San Lwin, Rohingya blogger; and Tun Khan, Chair of the Burmese Rohingya Organization of the UK.

6 The term refers to the sustainable use of ocean resources for economic growth, improved livelihoods and jobs, and ocean ecosystem health.

7 Personal interview with Shahidul Haque, former foreign secretary of the government of Bangladesh, February 27, 2019.

References

Abrar, Chowdhury R. 1996. "Voluntary Repatriation of Rohingya Refugees in Bangladesh." In *Proceedings of the Third Regional Consultation on Refugee and Migratory Movements*, edited by UNHCR. Colombo: UNHCR.

Abrar, C R. 2018. "Protected Return to Protected Homeland." *The Daily Star*. April 10. https://www.thedailystar.net/opinion/human-rights/protected-return-protected-homeland-1560379.

Association of Southeast Asian Nations. (n.d.). Charter of ASEAN. https://www.asean.org/wp-content/uploads/images/archive/21069.pdf.

BBC. 2018. "Aung San Suu Kyi: Amnesty Strips Myanmar Leader of Top Prize." 12 November 2018. https://www.bbc.com/news/world-asia-46179292.

Bhuiyan, Humayun Kabir. 2019. "Dhaka Displaced but Yet to Raise Rohingya Pushback Issue with Delhi." *Dhaka Tribune*, 21 January 2019. https://www.dhakatribune.com/bangladesh/foreign-affairs/2019/01/21/dhaka-displeased-but-yet-to-raise-rohingya-pushback-issue-with-delhi.

Bilsborrow, Richard E., A.S. Oberai, and Guy Standing. 1984. *Migration Surveys in Low Income Countries: Guidelines for Survey and Questionnaire Design.* London: Croom Helm and ILO.

Chan, Aye. 2005. "The Development of a Muslim Enclave in Arakan (Rakhine) State of Burma (Myanmar)." *SOAS Bulletin of Burma Research* 3 (2). https://www.soas.ac.uk/sbbr/editions/file64388.pdf.

Choudhury, Angshuman. 2018. "Why Are Myanmar's Neighbors Ignoring the Rohingya Crisis?" *The Diplomat*, 25 September 2018. https://thediplomat.com/2018/09/why-are-myanmars-neighbors-ignoring-the-rohingya-crisis/.

Encyclopedia of Burma. 1964. Vol 9. Rangoon: Government Printing House.

European Commission. 2020. "EU Allocates €31 Million for Bangladesh and Myanmar for Rohingya Crisis." European Civil Protection and Humanitarian Aid Operations. March 3, 2020. Last accessed May 15, 2021. https://ec.europa.eu/commission/presscorner/detail/en/IP_20_371.

European Council. 2019. "Myanmar/Burma: Council Prolongs Sanctions." Council of the EU, April 29, 2019. Last accessed September 9, 2021. https://www.consilium.europa.eu/en/press/press-releases/2019/04/29/myanmar-burma-council-prolongs-sanctions/.

Finer, Samuel E. 1962. *The Man on Horseback: The Role of Military in Politics*. London: Pall Mall Press.

FRC. n.d. "Our Mission." Last accessed September 9, 2021. https://freerohingyacoalition.org/en/our-mission/.

Human Rights Council. 2019a. "Detailed Findings of the Independent International Fact-Finding Mission on Myanmar." Human Rights Council, 42nd session, 16 September 2019. https://www.ohchr.org/Documents/HRBodies/HRCouncil/FFM-Myanmar/20190916/A_HRC_42_CRP.5.pdf.

———. 2019b. "Situation of Human Rights of Rohingya in Rakhine State, Myanmar." *Report of the United Nations High Commissioner for Human Rights*, 40th session, 11 March 2019. https://reliefweb.int/sites/reliefweb.int/files/resources/A_HRC_40_37.pdf.

Human Rights Watch. 1996. "Burma: The Rohingya Muslims: Ending a Cycle of Exodus?" *UNHCR*, 1 September 1996. Geneva: HRW. https://www.refworld.org/docid/3ae6a84a2.html.

———. 2013. "Burma: End 'Ethnic Cleansing' of Rohingya Muslims, Unpunished Crimes against Humanity, Humanitarian Crisis in Arakan State." Human Rights Watch. https://www.hrw.org/news/2013/04/22/burma-end-ethnic-cleansing-rohingya-muslims.

Huntington, Samuel P. 1968. *Political Order in Changing Societies*. New Haven: Yale University Press.

Hussain, Zarir. 2019. "India Deports Second Rohingya Group to Myanmar, More Expulsions Likely." Reuters, January 2, 2019. https://www.reuters.com/article/us-myanmar-rohingya-india-idUSKCN1OX0FE.

Islam, Arafatul. 2020. "Saudi Arabia Pressures Bangladesh to Issue Passports to Rohingya." *Deutsche Welle*, October 7, 2020. https://www.dw.com/en/saudi-arabia-wants-bangladesh-to-accept-rohingya/a-55187748.

Kamruzzaman, Md. 2019. "Bangladesh Seeks Safe Zone for Rohingya Refugees." Anadolu Agency, February 12, 2019. https://www.aa.com.tr/en/asia-pacific/bangladesh-seeks-safe-zone-for-rohingya-refugees/1390522.

———. 2020. "Argentinian Court Decision Brings Hope for Rohingya." Anadolu Agency, June 2, 2020. https://www.aa.com.tr/en/americas/argentinian-court-decision-brings-hope-for-rohingya/1861967.

Karen Human Rights Group. 2020. *Minorities under Threat, Diversity in Danger: Patterns of Systemic Discrimination in Southeast Myanmar.* November 2020. https://www.khrg.org/sites/khrg.org/files/report-docs/cmm_english_final_0_1.pdf.

Lewa, Chris. 2009. "North Arakan: An Open Prison for the Rohingya in Burma." *Forced Migration Review* 32. https://www.fmreview.org/statelessness/lewa.

Lewis, Simon, and Daphne Psaledakis. 2019. "U.S. Slaps Sanctions on Myanmar Military Chief over Rohingya Atrocities." Reuters, December 10, 2019. https://www.reuters.com/article/us-usa-myanmar-sanctions-idUSKBN1YE1XU.

Malik, Shahdeen. 2000. "Refugees and Migrants of Bangladesh: Looking Through a Historical Prism." In *On the Margin: Refugees, Migrants and Minorities*, edited by C R Abrar, RMMRU, 11–39.

Massey, Douglas S., Joaquin Arango, Graeme Hugo, Ali Kouaouci, Adela Pellegrino, and J. Edward Taylor. 1998. *World in Motion: Understanding International Migration at the End of Millennium.* Oxford: Clarendon Press.

MSF. 2002. "Ten Years for Rohingya Refugees: Past, Present and Future." Medicins Sans Frontieres. Last accessed September 9, 2021. https://www.msf.org/ten-years-rohingya-refugeespast-present-and-future-report-summary.

Nyunt Khin. 2016. *Naing Ngan Ei Ah Nauk Hpet Ta Gar Pauk Ka Pya Tha Na* [The Crisis from the Western Gate of Burma.] Rangoon: Pan Myo Ta Yar Press.

OHCHR. 2017. *Mission Report of OHCHR Rapid Response Mission to Cox's Bazar, Bangladesh.* September 13–24, 2017. Last accessed September 9, 2021. https://www.ohchr.org/Documents/Countries/MM/CXBMissionSummaryFindingsOctober2017.pdf.

Organization of Islamic Cooperation. 2017. "OIC Condemns Violence and Abuse of Rohingya in Myanmar." OIC, February 9, 2017. Last accessed September 9, 2021. https://www.oic-oci.org/topic/?t_id=13141&t_ref=5757&lan=en.

Pasquini, P. 2019. "Myanmar's Blood Gems Help Fund Rohingya Ethnic Cleansing." *The Washington Report on Middle East Affairs* 38 (1), 57.

Permanent Peoples' Tribunal Judgment. *State Crimes Allegedly Committed against the Rohinygas, Kachins and Other Groups.* September 2017, 1–70. University of Malaya, Malaysia. http://permanentpeoplestribunal.org/wp-content/uploads/2017/11/PPT-on-Myanmar-Judgment-FINAL.pdf.

Poling, Gregory B. 2014. "Separating Fact from Fiction about Myanmar's Rohingya." *Center for Strategic and International Studies.* February 13, 2014. Last accessed September 9, 2021. https://www.csis.org/analysis/separating-fact-fiction-about-myanmar's-rohingya.

Prasse-Freeman, E. 2017. "The Rohingya Crisis." *Anthropology Today* 33 (6): 1–2. https://doi.org/10.1111/1467-8322.12389.

R.C. 2018. "The Rohingya Crisis Bears All the Hallmarks of a Genocide." *The Economist,* 23 May 2018. https://www.economist.com/open-future/2018/05/23/the-rohingya-crisis-bears-all-the-hallmarks-of-a-genocide.

Reuters. 2018. "Holocaust Museum Rescinds Awards to Myanmar's Suu Kyi." March 8, 2018. https://www.reuters.com/article/us-usa-myanmar-award-idUSKCN1GJ2U0.

———. 2019. "Violent deaths of Venezuelans in Colombia on the rise: report." Reuters, 25 August. https://www.reuters.com/article/us-venezuela-migration-colombia-idUSKCN1VF0PE.

Rosenthal, Gert. 2019. *A Brief and Independent Inquiry into the Involvement of the United Nations in Myanmar from 2010 to 2018.* United Nations, May 29, 2019. https://www.un.org/sg/sites/www.un.org.sg/files/atoms/files/Myanmar%20Report%20-%20May%202019.pdf.

Sabbir, A., A. Al Mahmud, and A. Bilgin. 2022. "Ethnic Cleansing of Rohingya. From Ethnic Nationalism to Ethno-Religious Nationalism." *Conflict Studies Quarterly* 39: 81–91. https://doi.org/10.24193/csq.39.6.

Siddiqui, Tasneem. 2018. "Global Human Movement, Humanitarian Issues and Bangladesh Foreign Policy." In *Changing Global Dynamics: Bangladesh Foreign Policy,* edited by AKM Abdur Rahman, 111–36. Dhaka: BIISS and Pathak Shamabesh.

Singer, Paul. 1991. *Dinámica de la población y desarrollo: el papel del crecimiento demográfico en el desarrollo económico.* México: Siglo XXI.

Sorwar, Alam. 2019. "Infographic: Top Rohingya Hosting Countries." Anadolu Agency, August 24, 2019. https://www.aa.com.tr/en/asia-pacific/infographic-top-rohingya-hosting-countries/1563674.

Stern, Claudio. 1988. "Some Methodological Notes on the Study of Human Migration." In *International Migrations Today: Emerging Issues*, vol. 2, edited by Charles W. Stahl, 28–33. Paris: UNESCO.

UNHCR. *The 1951 Convention Relating to the Status of Refugees and Its 1967 Protocol.* Last accessed September 12, 2021. http://www.unhcr.org/about-us/background/4ec262df9/1951-convention-relating-status-refugees-its-1967-protocol.html.

United Nations. n.d. "United Nations Charter, Chapter XV: The Secretariat." https://www.un.org/en/about-us/un-charter/chapter-15.

UNOCHA. 2018. "Rohingya Crisis One Year On: Holding on to Hope." 28 August. https://unocha.exposure.co/rohingya-crisis-one-year-on-holding-on-to-hope.

UN News. 2018. "'No Other Conclusion,' Ethnic Cleansing of Rohingya in Myanmar Continues – Senior UN Rights Official." United Nations. 6 March. https://news.un.org/en/story/2018/03/1004232.

———. 2020. "Bangladesh Urges Greater International Action on Rohingya Status." United Nations. September 26, 2020. https://news.un.org/en/story/2020/09/1073882.

US Department of Treasury. 2018. *Treasury Sanctions Commanders of the Burmese Security Forces for Serious Human Rights Abuses.* 17 August. https://home.treasury.gov/news/press-releases/sm460.

Van de Berg, Stephanie. 2019. "Gambia Files Rohingya Genocide Case against Myanmar at World Court: Justice Minister." Reuters. 11 November 2019. https://www.reuters.com/article/us-myanmar-rohingya-world-court-idUSKBN1XL18S.

Yegar, Moshe. 1972. *The Muslims of Burma: A Study of a Minority Group.* Wiesbaden: Otto Harrassowitz.

Yunnan Daily. 2018. "Yunnan Energy Investment 'China Myanmar Energy Cooperation.'" April 18, 2018.

Zarni, Maung, and Natalie Brinham. 2019. *Essays on Myanmar's Genocide of Rohingyas.* Dhaka: C R Abrar.

Zarni, Maung, Natalie Brinham, and Alice Cowley. 2017. "An Evolution of Rohingya Persecution in Myanmar: From Strategic Embrace to Genocide." https://www.mei.edu/publications/evolution-rohingya-persecution-myanmar-strategic-embrace-genocide.

Zarni, Maung, and Alice Cowley. 2014. "The Slow-Burning Genocide of Myanmar's Rohingya." *Washington International Law Journal* 23 (3). https://digitalcommons.law.uw.edu/wilj/vol23/iss3/8.

7

West African Migration Regimes and the Externalization of EU Migration Management Policies

Leander Kandilige and Thomas Yeboah

Introduction

An estimated 281 million people worldwide are classified as migrants living outside their country of birth or origin (UNDESA 2020). Population growth, rising inequality, increasing connectivity, climate change, trade, and demographic imbalances suggest that this figure is likely to grow in the coming decades (UNDESA 2020). Migration provides enormous benefits and opportunities for migrants, their families, communities of origin, and host societies. However, it can also undermine progress and development, particularly when it is poorly regulated (Mallett 2018). In line with the recognition of the enormous potential and actual contribution of migration to development, there have been several attempts at the global, regional, and national levels aimed at establishing frameworks to realize the developmental benefits associated with migration and human mobility.

In Africa, attempts at establishing frameworks aimed at harnessing the developmental potential of migration are not new. For example, in West Africa, where intra- and intercountry migration is a central pillar of livelihood-building processes, the Economic Community of West African States (ECOWAS) established the Protocol on Free Movement of Persons, Right of Residence and Establishment in 1979 (Protocol A/P.1/5/79) as a way to promote free movement of goods, persons, and services (Awumbila, Teye, and Nikoi 2018). The overwhelming bulk of voluntary and involuntary migration in or from the region, encompassing farm laborers, traders, skilled workers, refugees, and internally displaced persons, has been essentially intraregional

(Awumbila et al. 2014; Adepoju 2015). For example, UNDESA data show that 71.8 percent of migrants from the West African region migrate to destinations within the region (UNDESA 2018). Voluntary migration within the subregion is shaped largely by real or perceived poverty, the pursuit of education and employment opportunities, and the desire for better livelihoods and income, structured along with a north-to-south movement from the landlocked and dry regions of the Sahel to the southern coast of West Africa (Kress 2006). Movement within and, in particular, to preferred destinations is shaped by factors such as common official languages, proximity, ethnic ties, and colonial legacy (Awumbila, Teye, and Nikoi 2018).

Much of the contemporary research and policy focus on migration in or from Africa has commonly focused attention on movement to Europe, at least in part, as a result of the emergence of the so-called European "migration crisis" that gained heightened interest in 2015. However, as Boswell (2003) and Düvell (2012) clearly demonstrate, attempts at curbing "unwanted" migration from third countries, especially in Africa, date back to the early 1990s. A two-pronged approach of deploying both control and preventive measures has dominated European policy and foreign relations discourse since the 1990s. Externalization control measures focus on origin and transit countries, including, firstly, "border control, measures to combat illegal migration, smuggling and trafficking, and capacity building of asylum systems and migration management in transit countries" (Boswell 2003, 622). Secondly, the EU adopted measures to return asylum seekers and "illegal" migrants by instituting readmission agreements with third countries. Prevention measures are based on a logic of changing the factors that "force or encourage migrants and refugees to travel to the European Union" (Boswell 2003, 624). To achieve this, a scheme to apparently address the so-called "root causes" of migration and refugee flows was proposed. As Boswell (2003, 624) notes, this particular approach was a rehash of debates that date back to the early 1980s. The scale of migration flows has increasingly become a contentious phenomenon in Europe, at least since the end of the Cold War, leading to further strengthening border control measures to restrict the flow of migrants into European territories. For example, in 2005, the European Agency for the Management of Operational Cooperation at the External Borders of the EU, also known as Frontex, became operational. The tightening of border controls has also been viewed in the context of fighting terrorism, particularly in the aftermath of September 11, 2001 (Léonard 2010). Accordingly, the need to control unwanted migration has moved from the EU border itself toward communities, regions, and countries of origin and transit (Vives 2017). The excessive attention paid to the so-called migration crisis as a problem of numbers and scale has tended to ignore the fact that this "migration crisis" or "boat migration issue" was just one dimension of a trajectory of multiple crises. Crawley et al. (2018) argue that the crisis started not in 2015 but in 2008 with the global financial crisis that led to the European financial and debt crisis associated with far-reaching austerity measures. The ways in which the "migration crisis" has been presented reflect and reinforce a particular way of thinking about the dynamics of migration. Media, political, and policy narratives across Europe increasingly spoke about an unprecedented arrival of migrants, a single coherent flow of people that unexpectedly resulted in pressure at the southern border of Europe. This view has consistently permeated EU–African migration regimes for several decades, to the extent that little interest is paid to the "back stories" of migrants; instead, the dominant narrative is often generalized and makes erroneous assumptions on how and why movement occurs (Crawley et al. 2018).

The overall implication has been a very narrow view of migration, leading to the establishment of several bilateral and multilateral agreements between the EU or individual European countries and African countries to restrict the number and scale of migration from Africa to Europe (Zanker 2019; Mouthaan 2019). As noted earlier, the growing presence of migration issues in EU–African relations increasingly extends to migrant-sending and transit countries across the entire African continent. Although concerns about migration in Europe were initially largely framed and limited to internal policy, mainly through the establishment of the free movement of labor under the Single European Act of 1988 (coming into force in 1992) and the creation of a borderless Europe spearheaded by the Schengen Agreement of 1985 (coming into force in 1995), these existed alongside a notion of an "external border" fortified through a paradigm of "remote control and externalization" and also a paradigm of "fortified, yet smart external borders through technology, digitization and biometrization" (Hess and Kasparek 2017, 60). This historical external dimension of European migration policy has persisted on the agenda of the EU and individual European countries.

Externalization here is seen as a process through which the European Union outsources a share of the control and management of migration and its borders to other parties or states beyond its own territorial boundaries. Externalization practices are increasingly becoming more prevalent and an integral part of the geopolitics of mobility, including the functioning of the EU as an "ordering actor" with various border control measures implemented widely as the external element of EU migration policy (Casas-Cortes, Cobarrubias, and Pickles 2016).

To what extent has the externalization of Europe's migration policies to Africa contributed to promoting or restricting migration within the continent itself? The application of EU externalization measures to African migration is significantly impacting both the EU's partnership with Africa and Africa's own migration and development policies (Idrissa 2019). Drawing on a review of the empirical literature, this chapter focuses attention on regional attempts at promoting migration in Africa and how these are aided or frustrated by EU–African relations, particularly the EU's externalized migration management policies. In doing so, we focus on the ECOWAS free movement protocol as an entry point to the wider debates about attempts to promote migration and regional integration in Africa. We further discuss EU policy responses to manage or control migration from Africa to European countries and their overall implications for regional integration and migration within the African continent. The central argument here is that migration policies of the EU toward Africa have contributed to restricting migration within the African context, thereby undermining the goal of free movement protocols that seek to promote intraregional mobility and socioeconomic development in West Africa.

The remainder of this chapter comprises two major sections: the first section outlines the attempts at promoting intraregional migration and integration through the ECOWAS free movement protocol debates about free movement in West Africa. The second section focuses on the EU–African migration governance agenda. Utilizing selected case studies, we outline the competing interests and power relations. We examine the externalization of the EU migration policy through specific examples of multilateral and bilateral migration agreements initiated by the EU and individual EU member states with West African countries in efforts to control or govern migration from Africa to Europe. We highlight the diverse implications of EU externalization policies, although our emphasis is on the indirect impact of how such measures inadvertently affect West African states' regional free movement objectives in recent years. Other examples are also drawn upon from other parts of the continent beyond West Africa.

Promoting Migration for Regional Integration and Development in West Africa: Toward a Free Movement Protocol

Attempts at promoting regional integration as a pivot to socioeconomic development in West Africa are well established. Following the demise of colonial rule in the 1950s and 1960s, there were discussions among African states to create a continental union that would strengthen the bargaining power of Africa at the global level, creating a continental federation and leveraging existing trade links and markets (Yeboah et al. 2020). In the early 1970s, leaders of the West African subregion further acknowledged the potential of regional integration as an essential primary step to the subregion's collective integration into the global capital economy (Adepoju 2009). While much recent discussion on regional integration attempts in West Africa is seen as mirroring European models, historical accounts suggest that regionalism in the subregion is fundamentally driven by local aspirations and realities: the need to harmonize, expand, and collectively leverage larger markets; political unity as a prelude to socioeconomic development, encouraging closer cultural ties, historical human, ideational and trade links; and expansion of educational attainment and outcomes (Yeboah et al. 2020). In this regard, regional integration in the subregion could be seen in the light of attempts to implement and realize a continental aspiration. The treaty establishing the Economic Community of West African States was signed in Lagos on May 28, 1975, by leaders of the following nations: Burkina Faso, Guinea-Bissau, Niger, Mauritania, Togo, Mali, Benin, Senegal, Côte d'Ivoire, Ghana, Guinea, Nigeria, Liberia, The Gambia, Sierra Leone, and Cape Verde. Article 27 of the treaty asserted a long-term goal to create a common community citizenship for nationals of all member states. A fundamental objective of the preamble to the treaty is to eliminate the barriers to the free movement of capital, goods, and persons in the subregion (Adepoju 2009). In furtherance to the commitments embedded in the protocol, in 1979, the protocol relating to the free movement of persons and the rights of residence and establishment was adopted and followed by several supplementary protocols in 1985, 1986, 1989, and 1990.

The protocol was intended to be implemented in three phases. Phase one, which focused on guaranteeing the right of entry to community citizens for an initial ninety-day stay without a visa, was ratified and implemented by member states in 1980. The second phase, the right of residence, enjoins member states to grant the right of residence to community citizens, stating the reference to compliance with established national procedures in carrying out these practices (Article 4, Protocol A/P/3/5/82 Relating to the Definition of Community Citizen). The protocol enjoins member states to ensure that migrant workers enjoy the same treatment as nationals in terms of access to sociocultural and health facilities and security of employment (ECOWAS Community 1982). The third phase of the protocol, the right of establishment, grants community citizens the right to settle in a member country and seek economic opportunities, like that of setting up and managing businesses in accordance with the legal framework of the host country for its own nationals. In addition, in 2008, member states adopted the ECOWAS Common Approach to migration. Issues discussed in the Common Approach were informed by the challenges and lived realities of the free movement protocol. Six main principles were outlined as guidelines for effective migration management: free movement of persons within the ECOWAS zone; promoting legal (regular) migration as an integral part of the development process; combating human trafficking; policy harmonization; protection of the rights of migrants, asylum seekers, and refugees; and recognition of the gender dimensions of migration. In 2000, the Authority of Heads of State and Government adopted a uniform ECOWAS passport, modeled on the EU passport and with the ECOWAS emblem on the front cover. A five-year transitional period was foreseen during which national passports would be used in conjunction with ECOWAS passports while ECOWAS passports were phased in and became more widely available (Adepoju 2009).

Several implementation challenges are associated with the ECOWAS free movement protocol, particularly with respect to phases two and three. Issues such as lack of coherence between member state laws and provisions in the protocol have impeded the smooth implementation of the second and third phases (Dick and Schraven 2018). Also, the procedures regarding immigration and emigration have not been fully harmonized, which also negatively impinges on data collection mechanisms. Monitoring the implementation of the protocol and the status of the free movement of persons in the region has also not been implemented by all member states. This is partly due to resource constraints. As of 2018, only 55 percent of the member states had established a national committee for monitoring the free movement of persons and goods (Butu 2013; Awumbila et al. 2014). In addition, only four out of fifteen member states have ratified the supplementary protocol,[1] which leaves room for varying interpretations of the protocols by the national authorities (Awumbila et al. 2014; Teye, Awumbila, and Benneh 2015).

Despite substantial implementation challenges, one cannot underestimate the extent to which the protocol has contributed to enhanced mobility, economic growth, and regional integration in the ECOWAS region. For instance, all member states have granted visa-free access for ninety days to eligible community citizens, enabling free movement to other member states (Awumbila et al. 2014). The implementation of the protocol has also yielded large economic benefits in terms of boosting intraregional trade and supporting the livelihood of community citizens (Awumbila, Teye, and Nikoi 2018).

For example, an estimated 74 percent of migrants in Nigeria are of ECOWAS member states, and around 60 percent of non-Ghanaian residents in Ghana are nationals from ECOWAS countries. These populations are building their livelihoods in the trade, oil and gas, industrial, and service subsectors in these countries. Their livelihoods contribute to income generation, asset accumulation, and increased remittances to their families in their home countries. In effect, despite the identified implementation challenges, the ECOWAS free movement protocol has contributed to enhanced mobility and, to some extent, regional integration and economic development in West Africa. It is important to note that discussions are currently underway to further regional integration and the ECOWAS protocol, including the following: the proposal to lift the provision that restricts member state citizens to enter and stay for a maximum of ninety days; the proposal for the establishment of a common social security; and the proposal for a common currency.

The promise of further and improved integration is at risk of being curtailed by the activities of the EU in its attempt to devolve external border management responsibilities to third countries in Africa, broadly, and more specifically in West Africa.

The EU–Africa Migration Governance Agenda: Differences in Interests and Power Relations

Even though migration within Africa and from Africa to Europe is not new, it is only in the recent past, at least since the 1990s, that the EU and individual European countries have been actively engaged in pursuing and establishing bilateral and multilateral migration frameworks with several migrant originating and transit countries in the Global South, particularly in Africa (Adepoju, Van Noorloos, and Zoomers 2010). Adepoju, Van Noorloos, and Zoomers (2010) highlight a chronology of EU externalization measures, with earlier bilateral and multilateral agreements focusing on controlling irregular migration and readmission agreements, followed by a period where agreements centered on labor migration. More recent attempts have focused on restricting movement as well as incorporating development aid as part of efforts to manage migration flows, for example, the "co-development" agreement between France and Mali (Adepoju, Van Noorloos, and Zoomers 2010). Many of the bilateral and multilateral agreements have addressed a wide range of issues, including promoting integration and return, family reunification, and social security, and preventing and combating irregular labor migration. They also seek to provide a mechanism for processes and organization of repatriation, especially of irregular African migrants in Europe. Many of the frameworks or agreements also aim to foster an "improved migration management mechanism" via capacity building, dialogue, and cooperation between the EU and the countries involved. Others have focused on the ethical recruitment of health professionals, as well as enhancing the opportunities for regular mobility, facilitated largely by labor market conditions or needs and demographic trends in several countries (Adepoju, Van Noorloos, and Zoomers 2010).

As noted by a growing body of research literature, many of the bilateral and multilateral migration agreements between the EU and African nation-states have essentially focused on initiatives to restrict and reduce the number and scale of irregular or "boat migration" from the region (Zanker 2019; Mouthaan 2019; Vives 2017; Dover 2008; Natter 2013; Crawley and Blitz 2018). Writing about the management of the "migration crisis" between West Africa and the EU, Vives (2017) argues that labeling the arrival of a small number of West African migrants to the Canary Islands between 2005 and 2010 as a "crisis" led to the development of a combination of preventive and defensive measures by the EU to buy the cooperation of West African nation-states in implementing more stringent migration and border controls outside the EU's territory. The defensive mechanisms operated through externalization of migration control responsibilities, militarization, and the return of and removal of undocumented migrants. On the other hand, Vives (2017) reports the creation of jobs in communities of origin, cooperation for development, and temporary migration programs as the key preventive measures. The combination of defensive and preventive measures is arguably mutually dependent. Their interplay is fundamental to understanding new forms of border territoriality in the context of migratory flows directed not just to Europe but also within and between countries in West Africa.

Essentially, the EU has adopted a harsh posture toward the arrival of migrants and the stay of irregular migrants from the sub-Saharan African region in member countries. Against the backdrop of the European "migration crisis" of 2015–16, the EU's approach to managing migration has intensified. As noted earlier, increasing attempts are being made to govern the external borders of Europe from outside of its own territory (Andersson 2014). Africa is a primary focus for Europe, as a source continent, in terms of Europe's fight against irregular migration. According to Mouthaan (2019), of the sixteen "priority countries" under the EU's Partnership Framework, thirteen are African countries.[2] Using the heavy financial muscle of the EU Emergency Trust Fund for Africa, managers of the EU's migration policies seem to have adopted a two-pronged approach: prevent migrants from arriving in the first instance and/or deport those who have succeeded in accessing the EU's territorial space irregularly. Slagter (2019) and Zanker (2019) note that the EU has changed tactics from relying on overt readmission agreements, struck with mostly developing countries of origin whose populations rely heavily on remittances, to more clandestine agreements that are mostly "out of sight" and evade domestic protestation and scrutiny. Slagter (2019) points to the interregional dialogues and return procedures initiated at the behest of the EU with their African counterparts, such as the Cotonou Agreement (signed in 2000), the Valletta Summit on Migration (2015), and the 5th African Union-European Union Summit in Abidjan (2017), which have all sought to lure third countries to readmit their nationals who are deemed to be in Europe irregularly without the need to adhere to strict deportation formalities. These are attempts to overcome earlier inconclusive negotiations with key source or transit countries such as Morocco (2000), Algeria (2002), and Nigeria (2016). Cape Verde is an exception, as a legally binding readmission agreement was concluded in 2013. Return of irregular migrants has focused the minds of European leaders irrespective of their ideological leanings (i.e., both left-wing and right-wing political parties). This rare, unified position is demonstrated, for instance, in the 2016 Partnership Framework, which overtly states that the "paramount priority is to achieve fast and operational returns," whether through formal readmission agreements or other informal arrangements (European Commission 2016a, 7; Mouthaan 2019).

A slew of informal agreements signed with some sub-Saharan African countries (e.g., Guinea in 2017; Ethiopia in 2018; The Gambia in 2018; Côte d'Ivoire in 2018) are, however, proving to have minimal effect on the actual numbers returned to these countries from the EU (Slagter 2019). In addition, individual EU countries such as the United Kingdom drafted memoranda of understanding (MoUs) with sub-Saharan African countries as a shortcut to returning undesirable African migrants. Though such MoUs were entered into with Nigeria (2005), Angola (2007), Somaliland (2007), Rwanda (2008), the Democratic Republic of the Congo (2009), Sierra Leone (2012), and South Sudan (2013), existing data suggests that there has been no discernible increase in deportations (Slagter 2019).

Some literature has argued that African governments are not "passive agents" in negotiating the EU's external migration governance arrangements despite a power imbalance between the two parties (Mouthaan 2019). This reasoning is attributed to the role of "local actors" inter alia political parties, media, civil society, and even the diaspora. While acknowledging the capacity of electoral pressures, media attention, public opinion, and diaspora influence in dissuading African governments from signing off on restrictive migration policies from the EU, this is only possible when the details of those agreements are made public. Clandestine and informal agreements or memoranda of understanding are not easily available for such scrutiny. The "blackmail" of the EU tying development aid and visa facilitation to African countries' willingness to take back their citizens who are deemed to be living in Europe irregularly is intended to deny source countries the agency to contest such policies (Den Hertog 2016). As such, we argue that African countries commonly acquiesce to European directives fundamentally as a result of economic and political power imbalances in favor of Europe.

More importantly, picking off the weakest links, such as Niger, and entrusting them with the responsibility of policing the EU's external territory against movement by fellow ECOWAS citizens who are presumed to automatically be destined for Europe is problematic. Circulation within ECOWAS is guaranteed by the ECOWAS protocol (Articles 2 and 27 on the abolition of visa requirements and conferment of community citizenship on all members of ECOWAS states, respectively). The receipt of financial assistance from the EU and the attendant quid pro quo obligation on Niger to disrupt the migration journeys of fellow West Africans puts Niger in jeopardy of its regional legal commitments as a signatory to the ECOWAS Free Movement Protocol. However, as the poorest country in Africa and one of the poorest globally, Niger is not able to resist the lure of financial assistance needed for the basic survival of its people. As such, irrespective of Niger's intentions to respect the spirit and letter of the ECOWAS protocol, successive governments have had to make pragmatic choices in favor of collaborating with the EU in the face of abject poverty. The EU's own records[3] note that its involvement has taken the form of equipment support to the Nigerien government, economic support to the city of Agadez, and the establishment of field offices in Agadez for capacity building and provision of direct budgetary support with the hope of dismantling the activities of the illegitimate actors in Niger's migration industry. Indeed, around 45 percent of the Nigerien government's annual budget is financed by external support, mainly including the EU's Emergency Trust Fund for Africa (EUTF), the World Bank, the United Nations, the African Development Bank, and other EU countries. The EUCAP Sahel Niger, a civilian EU Common Security and Defense Policy mission launched in August 2012, has seen its budget grow from €18.4 million to €63.4 million since it began. Faced with numerous developmental and security-related challenges, the Nigerien government has been extremely desperate to receive EU aid while being pressured to initiate anti-trafficking laws banning migrant transportation, but this has threatened local livelihoods and has thrown the economy into disarray (Creta 2020).

In addition, the declaration of migration as a crisis in 2015 led to the formulation of the Valletta Action Plan, in which Niger is the only West African country to be regarded as the key partner in the design of EU strategies in the region. Idrissa (2019) views the Valletta Action Plan as a "smorgasbord of projected initiatives" with a wide range of assistance covering support for private investment, job creation, scholarships for African students, and aid for refugees in camps. It emphasized the need to "establish and upgrade national and regional anti-smuggling and anti-trafficking legislation" through concerted action plans devised for individual countries based on their respective status as origin or transit countries. At their core, such plans focused on building the capacity of local security authorities to curb migrant transport or "human smuggling and trafficking." Under the same initiative, the EU devised what they called "effective incentives" with the view to develop legislative instruments that could enforce the adoption and implementation of its policies. Unlike Nigeria and Mali, where the adoption of migration policies in line with the EU action plan had essentially declarative rather than performative significance, Idrissa (2019) argues that, in 2015, the EU forced Niger to adopt the legislative instrument to criminalize migrant transportation. Quite unusually for a country that has a historically "lethargic rule of law apparatus," Niger soon started to implement this law in 2016, and the EU is reported to have awarded Niger funds to the tune of €640 million for the period 2014–20 and, after further appraisal, supplemented this with a further €95 million. Another support for Niger has come in the form of well-paid training sessions and receipt of equipment by Niger border control security forces; extension of electricity and internet connectivity to remote border posts; and donation of satellite phones, off-road vehicles, flatbed trucks, and motorcycles to improve the operations of the Niger's military and gendarmerie (Idrissa 2019).

What is regarded as "progress" is focused on reducing the number of migrants presumed to be using Niger as a transit country and reducing human trafficking and smuggling activity. To the extent that some of these activities/movements are actually targeted at arrivals to Europe, they may seem to be of legitimate interest to the EU. However, ostensibly, this is a blunt instrument, as it is incapable of distinguishing between ECOWAS citizens exercising their rights under the protocol and those who actually intend to proceed to Europe. Intervening opportunities in Libya, especially prior to the Libya crisis of 2011, meant that the majority of sub-Saharan African migrants actually sought employment in Libya rather than attempting to migrate to Europe. Therefore, the "sledgehammer" approach tends to treat all sub-Saharan African migrants (found in Niger) as collateral damage in an EU externalized migration governance system. Some research suggests a reduction of migrants transiting Niger to Libya from 400,000 recorded in 2016 to less than 10,000 in 2019 (Creta 2020). While the EU might have pushed Niger to crack down on the smuggling and trafficking of migrants, more dangerous routes have emerged, and the migrant exodus has drastically reduced. Official data from the International Organization for Migration indicate that migrant flows across Niger doubled from an estimated 266,590 in 2018 to more than 540,000 in 2019 (Creta 2020). The interception and return of some migrants south of the Niger border encompass those who had no intentions of journeying to Europe. What we see, then, is that Niger's decision to take over Europe's "dirty work" of border control and interception in the northern desert is a violation of the ECOWAS regional protocol on free movement, which the country is signatory to. Idrissa (2019) argues that in practice, Niger's agreement with the EU to take on border control measures and interception of migrants derived from the Valletta Action Plan and the Rabat Process appears to supplant the ECOWAS protocol in two ways. First, ECOWAS officials participated in the Rabat Process and willingly or otherwise endorsed the implementation of the policy measures adopted. Second, the ECOWAS guidelines are designed to provide gaps for the EU to configure them in ways that are favorable to their approach. The ECOWAS measures rest largely on member states establishing the architecture or infrastructure for border control to the extent that the region has become essentially borderless.

What appears abundantly clear is that the EU is now pushing West African states to uphold the letter of its laws in relation to border control, which has implications on endangering the status quo of what constitutes attempts at promoting regional integration in the subregion. When one looks critically at the prevailing migration patterns in West Africa and the current capacity of some individual ECOWAS countries, it becomes clear that instituting border control measures is both impractical and, more importantly, a potential source of regional instability or crisis. In this regard, what the EU–Niger agreement teaches us is that "the migration rules of ECOWAS now exist, in the Sahel, at least, under a regime largely defined by the concerns and interests of the EU" (Idrissa 2019, 35).

As Mouthaan (2019) argues, Senegal has more experience and leverage in negotiating agreements with the EU, having been designated as one of the five "compact" countries and one of the main beneficiaries of EUTF funding since the inception of the funding mechanism in 2015. Practical "support" to Senegal has been through funding of projects to bolster job creation targeted specifically at the youth; projects to build the capacity of Senegalese law enforcement agencies and border management officials; return and readmission programs; and enhancement of Senegal's ability to fight human smuggling. These pseudo-development support packages are narrowly focused on ensuring the immobility of African migrants, especially the youthful population, and the return of the "undesirables"—irregular African migrants who work in the informal economies of Europe doing the dirty, dangerous, and dehumanizing jobs that citizens of the host societies shun until African migrants became pawns in European anti-immigrant politics. Vives (2017) also discusses securitized operations by the EU or member states along the territorial waters of Europe as well as within the sovereign territories of West African countries such as Senegal in an attempt to stem the flow of migrants from sub-Saharan Africa. Border control operations such as Frontex's HERA operation and the Seahorse and West Sahel Operations have employed satellite technology for the surveillance of what are meant to be sovereign African territories alongside a physical presence of Spanish and EU security forces and Spanish private security companies (Vives 2017, 212). Under the auspices of the preventive "root causes" approach, similar funding agreements have been reached with other West African countries such as Nigeria and Mali.

Outside of West Africa, the EU's externalized migration control policies are equally inhibiting the realization of free movement across the African continent as envisaged in Article 6 of the Protocol to the Treaty Establishing the African Economic Community Relating to Free Movement of Persons, Right of Residence and Right of Establishment. Though African states are not unanimous in their embrace of the continental free movement protocol, bilateral agreements between the EU and some member states such as Libya, Ethiopia, and Sudan make it even more difficult to conjure a united front in promoting visa-free travel within Africa. Historically, EU–Libya relations have been frosty following the Pan Am Flight 103 aircraft bombing over Lockerbie, Scotland, in 1988. Libya was accused of being responsible for the bombing, and the United Nations imposed severe sanctions. However, in the 2000s, Libya's relations with Europe started to normalize as Libya was recruited as a critical partner in combating irregular migration from Africa to Europe. The enhanced cooperation, especially starting in 2003, was premised on fears that deliberately lax controls by Libya resulted in pressure at Europe's Mediterranean borders (Perrin 2008, 2). Italy, for instance, normalized its relations with Libya by signing multiple bilateral agreements on combating terrorism (in 2000), organized crime (in 2003), drug trafficking (2004), and irregular migration (2007). In addition, Italy and Libya signed a Friendship Pact in 2008, in which Italy agreed to pay Libya 5 billion US dollars over twenty years as reparation for colonization. As Hendow and Kandilige (forthcoming) note, the pact emphasized intensifying cooperation on combating irregular migration, which led to the practice of "pushbacks" from Italy to Libya (subsequently overturned by the European Court of Human Rights ruling, *Hirsi Jamaa and Others v. Italy* in 2012). Akin to later agreements with West African countries, the various Italian agreements with Libya embedded financial support for equipment and training programs for Libyan police and navy, construction of accommodation/detention centers for irregular migrants, payment for the return of migrants, and financing of repatriation flights as the quid pro quo for Libya serving as the ultimate gatekeeper of Europe's Mediterranean border.

The larger EU also sought Libya's assistance in combating irregular migration to Europe. Attempts took the form of a Memorandum of Understanding in 2007, followed by negotiations for a future Framework Agreement on Political Dialogue and Cooperation in 2008, and finally a cooperation agenda and Action Fiche in 2010 to the tune of €10 million covering irregular migration, border control and refugees (Hendow and Kandilige, forthcoming). In return, Gaddafi's government embarked on stringent measures to demonstrate their preparedness to address Europe's concerns about irregular migration routes from Libya. As Hendow and Kandilige (forthcoming) note, from the early 2000s, Gaddafi began to carry out large-scale expulsions of migrants (primarily sub-Saharan Africans), which grew from 4,000 people in 2000 to 43,000 in 2003 and over 64,000 in 2006 (Di Bartolomeo, Jaulin, and Perrin 2011; ICMPD 2010). Available data suggest that the most common nationalities apprehended and deported for irregular entry into Libya were Sudanese, Nigeriens, Chadians, Malians, Ghanaians, and Nigerians (ICMPD 2010). These actions contrasted sharply with earlier immigration policies by Libya, which were pro–sub-Saharan African migration: there were no visa requirements; only a medical certificate was required.

Following unsuccessful earlier attempts to halt irregular migration from Libya to Europe, in May 2015, the EU announced its military operations principally focused on the southern central Mediterranean. Its original name, EUNAVFOR-MED, was changed to Operation Sophia in September 2015, derived largely from the name of a baby born in one of the mission's ships off the coast of Libya. Operation Sophia had an overarching goal of "disrupting the business model of human smuggling and trafficking networks in the southern central Mediterranean" and was structured to be implemented in three phases (Baldwin-Edwards and Lutterbeck 2019, 2251). Phase one commenced with gathering intelligence on the activities of migrant networks, and phase two operations encompassed boarding, searching, arresting, and/or identifying vessels deemed to be used for the trafficking or smuggling of migrants on coastal waters. Phase three focused on disposing of vessels or other resources deployed for smuggling activities. In 2016, the activities of Operation Sophia were extended by a year, with two more tasks—reinforcing the implementation of the UN arms embargo on the high seas and training the Libyan coastguard to stem the flow of migrants from further south (Baldwin-Edwards and Lutterbeck 2019). In this regard, Operation Sophia, which had a simple mandate of surveillance and rescue mission, transformed into a full-blown Chapter VII operation under the UN Charter and was further extended until the end of 2018 under a decree by the EU Council in July 2017 (Blockmans 2016, 8). In 2017, Italy wrapped up a bilateral MoU with the newly constituted Libyan government headed by Fayez al-Sarraj, with full backing from the UN. The new agreement, which explicitly referred to the 2008 Friendship Treaty that existed between Italy and Libya, aimed at equipping the Libyan security force to stem irregular migration from along the country's southern borders. At the heart of the MoU was financial assistance to facilitate economic growth in spaces affected by irregular migration and to upgrade healthcare in migrant detention centers. Informed commentators, such as Toaldo (2017), argue that the MoU is flawed on human rights concerns, as it fails to establish an independent monitoring system or mention aspects of international conventions (only referring to International Customary Law).

Overall, by any standard of measuring policy effectiveness, Operation Sophia and other agreements between the EU and individual European countries, particularly Italy, and Libya have been a complete failure. Although such operations led to the interception of tens of thousands of people at sea, such militarized operations have been done in full awareness of the systematic violation of migrants' rights in Libya, including forms of rape and torture. Amnesty International (2020) data show that at least 40,000 persons, including thousands of children, were intercepted at sea and returned to Libya, but such people were exposed to horrific suffering. Indeed, such migrants are unlawfully detained and face severe risks of abuse, torture, rape, and overcrowding (Amnesty International 2020). Moreover, official statistics from UNHCR (2016) reveal that arrivals in Italy through the central Mediterranean route in real terms increased to reach a record high of 180,000 in 2016 after Operation Sophia was launched.

Additionally, the number of migrant deaths stood at around 5,000 in 2016. Smugglers have been reported to avoid being on boats, preferring instead to offer free passage to young minors to guide the vessels into international waters until they are recognized and rescued (Baldwin-Edwards and Lutterbeck 2019, 2251). Apart from the growing fatalities and abuse of the human rights of migrants, we argue that using the reduction in the number of migrants arriving in Libya from countries further to the south as a measure of the success of EU externalization policies indirectly contributes to restricting free movement of persons on the African continent. In fact, some research findings suggest that some African migrants arriving in Libya have no aspirations or intentions of onward journey to Europe (Crawley et al. 2018).

In addition to stepping up its patrol activities in the Mediterranean and seeking to strengthen the (re)engagement with Libya in combating irregular migration, the EU externalization control measures have been extended to countries further south along the central Mediterranean migration route, such as Niger and Mali, in order to prevent irregular migration across Libya's southern borders. In 2011, a new policy, Strategy for Security and Development in the Sahel, with an initial focus on alleviating a number of developments and security-related challenges in the Sahel region, was initiated shortly after the Libyan uprising. It aimed to combat rapid population growth, the effects of climate change, corruption, illicit trafficking, and terrorism-related security threats, unresolved internal tensions, fragile governance, the risk of violent extremism and radicalization, extreme poverty, and frequent food crises (Baldwin-Edwards and Lutterbeck 2019). Two key strategies, including the launch of a capacity-building program and the strengthening of regional cooperation in areas of government activity between the Sahel countries, were rolled out to achieve the agenda. In 2012, following the request of the Niger government, the EUCAP Sahel Niger mission, which was the first-ever mission under the Sahel strategy, was launched. It aimed "to develop an integrated, coherent, sustainable and human rights-based approach among the various Nigerien security agencies in the fight against terrorism and organized crime by providing training and advice to Nigerien authorities" (Baldwin-Edwards and Lutterbeck 2019, 2251). However, what is abundantly clear is the fact that the EU's Sahel strategy has taken an explicit focus on stemming irregular migration. There is a sharp overlap between the EU's "Partnership Framework" approach under the European Agenda on Migration and the Sahel strategy. For example, in

2014, the overarching objectives of the EUCAP Sahel Niger mission were revised, whereby curbing irregular migration from countries further south via Niger to Libya became the main mission objective (EEAS 2016). A permanent field office was set up in Agadez, whose main task was to provide training to Nigerien security forces and authorities in areas such as criminal investigation methodology, trafficking in human beings, and document forgery (European Commission 2016b; Baldwin-Edwards and Lutterbeck 2019). In 2017, these efforts were further scaled up with the deployment of the EU Migration Liaison Officer in Niamey. Under the partnership policy framework, several other countries along the central Mediterranean migration route, including Senegal, Ethiopia, and Mali, were identified as priority partners. In 2015, a comparable EUCAP mission was inaugurated in Mali. Furthermore, while the stated decree mission has been to enhance the overall efficacy of the security sector of the country, in practice, assistance has been to consolidate the capacity of the state to control its borders and stem migration flows from neighboring West African countries (European Commission 2017; Baldwin-Edwards and Lutterbeck 2019).

Conclusion

Promoting free movement of persons, goods, and services has long been on the agenda of West African states, and the coming into effect of the ECOWAS free movement protocol marks an important advance in the commitment of ECOWAS member states to strengthen regional integration and mainstream the potentials of intraregional migration for development. We have established that the implementation of free movement protocols and frameworks has, to a large extent, contributed to enhanced mobility, regional integration, and socioeconomic progress within West Africa. However, the quest to fight against "undocumented migration" from West Africa to Europe has impelled a broadening of activities aimed at protecting the EU's external borders and transforming the relationship between the European Union and West African nation-states in migration management and control. There has been the establishment of several bilateral and multilateral agreements between the EU and/or individual EU countries on the one hand and African countries on the other. Such agreements and frameworks usually aim at combating irregular or "boat migration," promoting integration and return, improving migration management mechanisms, and enhancing opportunities for regular migration to Europe. We argue that there is an important degree of power asymmetry in developing and implementing such agreements. In most cases, it is the agenda of the EU that is spearheaded, often with little or no agency of the African country involved. Thus, EU external migration policies have been deployed to buy the cooperation of West African nation-states in implementing more stringent migration and border controls outside the EU's territory, but at the expense of free mobility initiatives within the West African subregion.

In conclusion, we are of the view that the continued presence and involvement of the EU and individual EU countries in establishing bilateral or multilateral agreements have negatively impacted prospects for free movement, regional integration, and socioeconomic development, which have long been important goals of West African nation-states. Such agreements have further contributed to the loss of human life and the criminalization of migration and exposed many migrants, including children, to abuse, rape, and torture in detention centers. While not being overly preemptive, we foresee that the continuous establishment of these more stringent migration and border controls within African countries on behalf of the European Union would also negatively impact the continental free trade agreement of the African Union that seeks to establish and create a single continental market for goods and services, with free movement of business, persons, and investments.

Notes

1 The Supplementary Protocol A/SP.1/7/86 is on the Second Phase (Right of Residence). Article 2 specifically requires member states to grant community citizens who are nationals of other member states the right of residence in its territory for the purpose of seeking and carrying out income-earning employment after obtaining a residence card or permit.

Supplementary Protocol A/SP.2/5/90 is on the implementation of the Third Phase of the Free Movement Protocol (Right of Establishment). The protocol emphasizes the non-discriminatory treatment of nationals and companies of other member states except as justified by exigencies of public order, security, or health (Articles 2 to 4).

2 Ethiopia, Eritrea, Mali, Niger, Nigeria, Senegal, Somalia, Sudan, Ghana, Côte d'Ivoire, Algeria, Morocco, Tunisia, Afghanistan, Bangladesh, and Pakistan.

3 Niger action and progress under the migration partnership framework, June-December 2016.

References

Adepoju, Aderanti. 2009. "Creating a Borderless West Africa: Constraints and Prospects for Intra-regional Migration." In *Migration Without Borders: Essays on the Free Movement of People*, edited by Antoine Pécoud and Paul de Guchteneir, 161–74. Paris and New York: UNESCO Publishing and Berghahn Books.

———. 2015. "Operationalizing the ECOWAS Protocol on Free Movement of Persons: Prospects for Sub-regional Trade and Development." In *The Palgrave Handbook of*

International Labor Migration, edited by Marion Panizzon, Gottfried Zürcher, and Elisa Fornalé, 441–62. London: Palgrave Macmillan.
Adepoju, Aderanti, Femke Van Noorloos, and Annelies Zoomers. 2010. "Europe's Migration Agreements with Migrant-sending Countries in the Global South: A Critical Review." *International Migration* 48 (3): 42–75.
Amnesty International. 2020. "Libya: Renewal of Migration Deal Confirms Italy's Complicity in Torture of Migrants and Refugees." January 30, 2020. https://www.amnesty.org/en/latest/news/2020/01/libya-renewal-of-migration-deal-confirms-italys-complicity-in-torture-of-migrants-and-refugees/.
Andersson, Ruben. 2014. *Illegality, Inc.: Clandestine Migration and the Business of Bordering Europe*. Oakland: University of California Press.
Awumbila, Mariama, Yaw Benneh, Joseph Kofi Teye, and George Atiim. 2014. *Across Artificial Borders: An Assessment of Labour Migration in the ECOWAS Region*. IOM: UN Migration. Brussels: ACP Observatory on Migration. https://publications.iom.int/books/across-artificial-borders-assessment-labour-migration-ecowas-region.
Awumbila, Mariama, Joseph Teye, and Ebenezer Nikoi. 2018. *Assessment of the Implementation of the ECOWAS Free Movement Protocol in Ghana and Sierra Leone*. MADE: Migration and Development Civil Society Network. https://www.madenetwork.org/sites/default/files/CMS%20research%20Guinea%20Sierra%20Leone%20WA%202018.pdf.
Baldwin-Edwards, Martin, and Derek Lutterbeck. 2019. "Coping with the Libyan Migration Crisis." *Journal of Ethnic and Migration Studies* 45 (12): 2241–57. doi: 10.1080/1369183X.2018.1468391.
Blockmans, Steven. 2016. "New Thrust for the CSDP from the Refugee and Migrant Crisis." *CEPS Special Report*, 142. July 2016. http://aei.pitt.edu/78095/1/Thrust_to_CSDP_S_Blockmans_CEPS_Special_Report.pdf.
Boswell, Christina. 2003. "The 'External Dimension' of EU Immigration and Asylum Policy." *International Affairs* 79 (3): 619–38.
Butu, Ali Williams. 2013. "Impact of ECOWAS Protocols on Political and Economic Integration of the West African Sub-region." *International Journal of Physical and Human Geography* 1 (2): 47–58.
Casas-Cortes, Maribel, Sebastian Cobarrubias, and John Pickles. 2016. "'Good Neighbors Make Good Fences': Seahorse Operations, Border Externalization and Extra-territoriality." *European Urban and Regional Studies* 23 (3): 231–51.
Crawley, Heaven, and Brad K. Blitz. 2018. "Common Agenda or Europe's Agenda? International Protection, Human Rights and Migration from the Horn of Africa." *Journal of Ethnic and Migration Studies* 25 (12), 2258–74. doi: 10.1080/1369183X.2018.1468393.
Crawley, Heaven, Franck Düvell, Katharine Jones, Simon McMahon, and Nando Sigona. 2018. *Unravelling Europe's "Migration Crisis": Journeys over Land and Sea.* Bristol, UK: Policy Press.
Creta, Sara. 2020. "How Niger Became Europe's Invisible New Border." RTE, February 12, 2020. https://www.rte.ie/brainstorm/2020/0212/1114761-niger-europe-border-eu-migration/.
Den Hertog, Leonhard. 2016. "EU Budgetary Responses to the 'Refugee Crisis': Reconfiguring the Funding Landscape." *CEPS Paper in Liberty and Security in*

Europe 93. Retrieved from SSRN: https://ssrn.com/abrastract=2786930.

Di Bartolomeo, Anna, Thibaut Jaulin, and Delphine Perrin. 2011. *CARIM–Migration Profile: Niger.* https://cadmus.eui.eu/bitstream/handle/1814/22442/migration%20profile%20FR%20Niger%20-%20links.pdf?sequence=2.

Dick, Eva, and Benjamin Schraven. 2018. "Regional Migration Governance in Africa and Beyond: A Framework of Analysis." *Deutsches Institut für Entwicklungspolitik (DIE).* German Development Institute. https://www.idos-research.de/en/discussion-paper/article/regional-migration-governance-in-africa-and-beyond-a-framework-of-analysis/.

Dover, Robert. 2008. "Towards a Common EU Immigration Policy: A Securitization Too Far." *European Integration* 30 (1): 113–30. https://doi.org/10.1080/07036330801959523.

Düvell, Franck. 2012. "Transit Migration: A Blurred and Politicized Concept." *Population, Space and Place* 18 (4): 415–27.

ECOWAS Community. 1982. *Protocol A/P.3/5/82 Relating to the Definition of Community Citizen.* May 29, 1982. http://citizenshiprightsafrica.org/wp-content/uploads/2016/06/ECOWAS-Protocol-Relating-to-the-Definition-of-Community-Citizen.pdf.

EEAS (European External Action Service). 2016. *Common Security and Defense Policy: The EUCAP Sahel Mali Civilian Mission.* Factsheet, June 2016. European External Action Service. https://www.eeas.europa.eu/eucap-sahel-mali/about-eucap-sahel-mali_en?s=331.

European Commission. 2016a. *Communication from the Commission to the European Parliament, the European Council, the Council and the European Investment Bank: On Establishing a New Partnership Framework with Third Countries under the European Agenda on Migration (COM/2016/0385 final).* Retrieved from https://eur-lex.europa.eu/legal-content/EN/TXT/?uri=CELEX%3A52016DC0385.

———. 2016b. *First Progress Report on the Partnership Framework with Third Countries under the European Agenda for Migration.* The European Commission, October 20, 2016. https://reliefweb.int/report/world/first-progress-report-partnership-framework-third-countries-under-european-agenda.

———. 2017. *Joint Communication to the European Parliament, the European Council and the Council: Migration on the Central Mediterranean Route. Managing Flows, Saving Lives,* 4. January 25, 2017. Brussels. https://eur-lex.europa.eu/legal-content/EN/TXT/?uri=CELEX%3A52017JC0004.

Hendow, Maegan, and Leander Kandilige. Forthcoming. "Violence against African Nationals during Libya's 2011 Crisis." In *When Migrants Are Caught in Crisis: A Comparative Perspective,* edited by Albert Kraler, Robtel Neajai Pailey, and Maegan Hendow. London: Routledge.

Hess, Sabine, and Bernd Kasparek. 2017. "Under Control? Or Border (as) Conflict: Reflections on the European Border Regime." *Social Inclusion* 5 (3): 58–68. https://doi.org/10.17645/si.v5i3.1004.

ICMPD. 2010. *A Comprehensive Survey of Migration Flows and Institutional Capabilities in Libya.* Vienna: ICMPD.

Idrissa, Abdoulaye. 2019. *Dialogue in Divergence: The Impact of EU Migration Policy on West African Integration: The Cases of Nigeria, Mali, and Niger.* Berlin: Friedrich Ebert Stiftung.

Kress, Brad. 2006. "Burkina Faso: Testing the Tradition of Circular Migration. Migration Information Source." *Migration Policy Institute*, May 1, 2006. https://www.migrationpolicy.org/article/burkina-faso-testing-tradition-circular-migration.

Léonard, Sarah. 2010. "EU Border Security and Migration into the European Union: FRONTEX and Securitization through Practices." *European Security* 19 (2): 231–54.

Mallett, Robert. 2018. "Decent Work, Migration and the 2030 Agenda for Sustainable Development." *Briefing Note*. Available at http://cdn-odi-production.s3-website-eu-west-1.amazonaws.com/media/documents/12422.pdf.

Mouthaan, Melissa. 2019. "Unpacking Domestic Preferences in the Policy-'receiving' State: the EU's Migration Cooperation with Senegal and Ghana." *Comparative Migration Studies* 7 (1): 1–20.

Natter, Katharina. 2013. "The Formation of Morocco's Policy towards Irregular Migration (2000–2007): Political Rationale and Policy Processes." *International Migration* 52: 15–28. https://doi.org/10.1111/imig.12114.

Perrin, Delphine. 2008. "Aspects juridiques de la migration circulaire dans l'espace Euro-Méditerranéen. Le cas de la Libye." [Migration Policy Centre], [CARIM-South], CARIM Analytic and Synthetic Notes, 2008/23, Circular Migration Series, Legal Module. https://hdl.handle.net/1814/8344.

Slagter, Jonathan. 2019. "An 'Informal' Turn in the European Union's Migrant Returns Policy towards Sub-Saharan Africa." Migration Policy Institute, January 10, 2019. https://www.migrationpolicy.org/article/eu-migrant-returns-policy-towards-sub-saharan-Africa.

Teye, Joseph Kofi, Mariama Awumbila, and Yaw Benneh. 2015. "Intra-regional Migration in the ECOWAS Region: Trends and Emerging Challenges." In *Migration and Civil Society as Development Drivers - A Regional Perspective*, edited by A.B. Akoutou, R. Sohn, M. Vogl, and D. Yeboah, 97–124. Bonn: Centre for European Integration Studies.

Toaldo, Mattia. 2017. "The EU Deal with Libya on Migration: A Question of Fairness and Effectiveness." Commentary, *European Council on Foreign Relations*. https://ecfr.eu/article/commentary_the_eu_deal_with_libya_on_migration_a_question_of_fairness_a/.

UNDESA. 2018. *International Migrant Stock*. United Nations: Geneva.

———. 2020. *International Migrant Stock*. United Nations: Geneva.

UNHCR. 2016. "Italy: UNHCR Update No. 10." December 2016. https://reliefweb.int/report/italy/italy-sea-arrivals-unhcr-update-10-december-2016.

Vives, Luna. 2017. "The European Union–West African Sea Border: Anti-immigration Strategies and Territoriality." *European Urban and Regional Studies* 24 (2): 209–24.

Yeboah, Thomas, Leander Kandilige, Amanda Bisong, Faisal Garba, and Joseph Kofi Teye. 2020. "Same Policy but Different Implications for Different People? An Analysis of How Diverse Categories of ECOWAS Nationals Experience the Free Movement Protocol." The Merian Institute for Advanced Studies in Africa (MIASA), MIASA Working Paper. University of Ghana. https://www.ug.edu.gh/mias-africa/sites/mias-africa/files/images/MIASA%20WP_12020.pdf.

Zanker, Franzisca. 2019. "Managing or Restricting Movement? Diverging Approaches of African and European Migration Governance." *Comparative Migration Studies* 7 (17): 1–18.

PART THREE
MIGRANTS, IM/MOBILITY, AND MIGRATION REGIMES

8

PRAN WOUT LA: Experiences and Dynamics of Haitian Mobility

Mélanie Montinard

Introduction

This paper examines the experiences and dynamics of Haitian mobility based on the narratives collected during my research (Montinard 2019).[1] It explores the different *wout* ("routes/roads/ways" in Haitian Creole) that Haitians have undertaken, specifically from Brazil, to *chache lavi* "pursue life" in their quest for *lavi miyò* "a better life," making the term *wout* a native category constitutive of Haitian mobility.

Since 2010, Brazil has become a transit place for numerous Haitian migrants seeking to reach French Guiana (Joseph 2015a). In the following years, the country thus became part of the Haitian sociogeographic space and a place of passage and residence[2] inserted within the broader repertoire of landscapes that map the diverse *wout* undertaken by Haitians to *chache lavi*. However, it is also important to consider Haitian society through its diaspora (Joseph 2015a, 2015b; Glick-Schiller 2011), recognizing there is no singular event that produced or provoked migration but rather a series of historical departures and migratory flows. Indeed, these comprise a set of different economic factors (around 59 percent of the population lives on an income below $2.41 per day according to a World Bank household survey conducted in 2012); social, cultural, and political factors (the authoritarian regimes of François Duvalier and Jean-Claude Duvalier from 1957 to 1986 and the political crises that followed: see Zolberg, Suhrke, and Aguayo 1989); and natural factors (such as floods, hurricanes, and the January 2010 earthquake) all intertwine to make Haiti[3] a country constructed and reconstructed through departures and mobility.

Moreover, we know from studies of the historicity of Haitian migration (Anglade 1982; Joseph 2015a) that for more than a century, Haitians have been attempting to reach the United States, one of the most favored destinations, to escape the economic and political problems at home. This migratory movement was made more intense and complex due to the economic recession in Brazil from the end of 2015, which directly affected those Haitians living in the country since 2010/2011, principally those who had benefited from the strong demand for workers due to international events such as the World Cup in 2014 and the Olympic Games in 2016. In this context, new migratory projects had seen the light of day, like the *wout Miami* (Miami road) or *pran wout la* (take to the road), expressions used by Haitians to refer to someone who has set off to reach the United States, traversing the borders of diverse South American and Central American countries to realize the dream of becoming *dyaspora*.

As I demonstrated in my doctoral research (Montinard 2019), *pran wout la* cannot be reduced to a single voyage from Brazil to the United States, a specific decision, or a specific itinerary featuring one or more countries, including the United States, Haiti, Brazil, Chile, Mexico, the Dominican Republic, or Canada, among others. The expression denotes both pragmatics of mobility and a state of being and becoming within the contemporary Haitian diaspora, making the term *wout* a native category for the purposes of comprehending the dynamics of the network of mobility:

> *Pran wout la* is a becoming, a state of construction, a mode of being in movement that includes physical and symbolic dimensions and can take constitutive derivatives of mobility (like *ouvè wout la* – "open the road," or *kite wout la* – "abandon the road," among others), as a way of being in different spaces and moments of mobility. (Montinard 2019, 257)

However, analyzing the native category *wout* through its terms and contexts within individual trajectories in perpetual motion, exploring the different relations and the (im)balances of logics of mobility, was as much a challenge for ethnographic study as it was a rich analytic source for an ethnographic theory of mobility. Indeed, the ethnographic research unfolded in multiple places and in movement, which posed unique methodological challenges, only made possible because of my multiple connections to Haitian networks and my immersion in migration public policy networks and debates. Thus, my research necessarily has a self-analytic or autoethnographic dimension, which forced me to consider the different commitments and places I occupy and the way in which my interlocutors and other actors perceive me. In this sense and within my ethnographic experience, the dimension of "participatory objectification" is indissoluble from that of "observant participation" (Bourdieu 1991, 2003).

During my field research between Haiti, Brazil, and the United States, I was able to observe the meanings of the word *wout*, a term used by Haitians either on the move or those preparing for a (new) departure. In fact, the word *wout* also has physical and symbolic dimensions. Toward the end of 2015, the expression *wout Miami* appeared in the everyday lives of those Haitians in Brazil who had set off for the United States. The term *wout* also became part of the vocabulary of Haitian mobility via the expression *pran wout la* (taking the road), blurring the boundaries between these separate dimensions. Inquiring into the meanings and uses of the native concept *wout* thus entails connecting the pragmatics of mobility to the pragmatics of the diaspora described by Joseph (2015a, 2015b) and Glick-Schiller (2011).

Derivations of the Word *Wout*

In Haiti, the word *wout* is mainly conjugated with the verb *faire* in the future tense, *m pral fè wout la, ann fè wout la* (literally I will do/make the route/road/way, set off on the road), which may reveal a physical dimension, a movement from one point to another, one place to another, signifying the route that someone takes to reach a desired destination.

Pran wout la (taking the road) is not only the journey, planned or not, undertaken by a person that becomes established in the duration, shorter or longer, between departure, travel, and arrival, between staying, leaving, and leaving again, between individual and collective strategies, navigating between the physical, symbolic, and collective dimension of the *wout*. The term also relates to an idea of privilege for those on the *wout*, to a process of success and construction of the meanings of the word *diaspora* despite having suffered *(pase mizè)*. Success is not always defined by obtaining a job or making a good living. It may have other meanings: for example, when a father brings his wife and son to live with him in Brazil, when two people marry in a church, when a child is born in Brazil, when someone arrives in Miami, or when someone is promoted to a managerial position at the Livraria Francesa in Brasilia—numerous possibilities that denote something other than the purely economic sense of success.

Suppose *pati* (to leave), *vwayaje* (to travel), *ale* (to go)[4] are verbs combining and interweaving the different meanings of Haitian mobility (Montinard

2019, 177–78). In that case, they reflect a social reality whose destination, whether final or not, tells us something about the representations of the category diaspora *(ti dyaspora, gwo dyaspora)*, as Joseph (2015b) has described and reflects the idea of a process of "becoming a diaspora." If *pati kite Ayiti* (to leave, to quit Haiti) has been the experience of thousands of Haitians,[5] the vast majority would repeat the words that Yves once declared to me: *Ayiti pap ka kite nou, li nan kè nou* (We cannot ever leave Haiti. It remains in our hearts), a statement that ended with a heavy sigh, as if Haiti, despite all its precariousness and lack of opportunities, represented, in the everyday lives of those who live abroad, a space of references, obligations, remembrance, and tensions where feelings of belonging mingle with memories and longing for the homeland. These ties are lived at a distance and (re)woven in different forms, during trips and visits made whenever possible, through money transfers to family, etc. However, Yves's sigh was related to an ambivalence between the fact that the *dyaspora* person never actually leaves Haiti and the fact that the country lives and breathes through the mobility of its people and represents a place that one must leave to *chache lavi*, not just for oneself but also for those left behind.

Thus, the *dyaspora* person is obliged to *pran wout la* (take the road) to rebuild a life abroad and seek to make it full and worthwhile (Neiburg 2017, 2019), meeting the moral obligations of the family and community, a reality that applies not only in Haiti but also to the country of passage or residence, such as in the case of Brazil.

To analyze the dynamics of networks and Haitian mobility to become a *dyaspora*, the rest of this paper traces the experience of Pipo's *wout*, a friend, who gave me the finer details of his journey. When Pipo announced that he was going to take the *wout Miami*, he justified it to himself by explaining: *bagay yo pa bon mwen bò isit, m gen yon pitit Ayiti* (things aren't good for me here, I have a daughter in Haiti), as though *chache lavi* in Brazil did not represent, or perhaps no longer represented, the meaning of being *dyaspora*. Pipo had to take a new *wout* to help his family back in Haiti, thus revealing that the outcomes of his journey to Brazil were a type of *wout* failure.

This situation reflects the plurality not only of the relations between individuals but equally between them and their place of residence or origin, revealing the perspectives, actions, and moral principles in play in the dynamics of mobility and the forms through which the phenomenon is constructed or deconstructed in people's everyday lives, mixing relational, emotional, and territorial proximities and distances in which the threat of frustration (*pwoblèm, fristrasyon, konfli*) remains ever present (Neiburg 2017).

Hence, the *wout Miami* corresponds (or corresponded) above all to representations of being (in the) diaspora, being able to send money to one's family regularly (in this case, in US dollars, a strong currency), being able to plan to bring a relative, or visiting Haiti, for example. These representations, this dream of becoming diaspora, had the consequence of opening the road *(ouvè wout la)* for thousands of Haitians living in Brazil.

When *Pran Wout La* Leads to Miami

It was a December evening in 2015, in the streets of Cidade de Deus, Rio de Janeiro, that I first heard the plans for new routes setting out from Brazil. I was immersed in a conversation in which Pipo was explaining to Bob, my husband, that he was leaving the next morning to *pran wout Miami*:

> Bob, you know how much I respect your leadership among Haitians in Rio de Janeiro. We are rastafari, you have taught us a lot, our *baz* [base][6] would no longer exist had you not been there. Out of respect, I need to tell you something: tomorrow morning, Etienne and I are leaving for São Paulo to take the *wout Miami*. We're going to try our luck.

This news about his imminent departure left me perplexed since I knew that thousands of other Haitians would soon set out on the same route *(vide sou wout la)*. Indeed, the consequences of the Haitian presence in Brazil, considered a government "problem" (Dias and Vieira 2019), had entailed the adoption of specific public policies to make this new population governable, thus provoking further academic debates about "governmentality" (Foucault 2004).

A few weeks before Pipo announced to us the news of his departure, the Brazilian government had published its decision in the *Federal Gazette* to regularize the legal status of nearly 44,000 Haitians, the vast majority of whom had entered across the Amazonian borders.[7] Although the economic recession had started to be felt in Brazil, most of the Haitians at the end of 2015 were formally employed, a situation revealed in the annual report for 2016 produced by the International Migration Observatory (OBMigra) in partnership with the Ministry of Labor and Employment and the University of Brasilia. The report showed that the Haitian population was the largest foreign nationality in the formal labor market, ahead of the Portuguese: the Haitians had increased from 815 immigrants in 2011 to 33,154 in 2015 (Cavalcanti, Riberio de Oliveira, and Araújo 2016). According to the report, in 2013, the Portuguese had already been overtaken by the Haitians, who then, in 2015, represented 26.4 percent of the immigrant labor force in

Brazil. While the situation by the end of 2015 suggested a positive balance sheet in terms of a certain willingness to integrate Haitian nationals in Brazil, whose public policies and economic situation were proving rather favorable, new routes were already emerging within Haitian networks in response to this imaginary where *chache lavi* was only possible on the lands of Uncle Sam.

Pipo told me about the individual and collective strategies he was mobilizing to prepare for his departure during the conversation. Although he had just brought one of his younger brothers to Rio de Janeiro, he decided to undertake the *wout Miami* with two childhood friends, Etienne and Kenken. The three friends had negotiated the termination of their contracts with the civil construction firm for which they worked and had used the redundancy money to organize their departure and fund their *wout* to Ecuador. Leaving behind his brother, Pipo would travel to São Paulo with his two friends, where they would spend some weeks before flying on to Rio Branco, Acre, in early February 2016.

Pipo's stories also revealed the physical and symbolic dimensions of the *wout Miami*, mobilizing personal and collective resources, strategies, encounters, chance and fate, the *lwa* (spirits) or the hopes in *Bondye* (God), when obstacles like mountains appeared ahead and where borders become not only physical but symbolic. On one of my trips to Florida, when I visited him in Margate, Pipo told me the details of his experience on the *wout Miami*. A conversation that began in a very serious tone:

> Everyone has experienced this *wout* in a different way. For me, this was a trip, you know. I'm a Rasta, I must approach each stage of life in a positive way and believe in myself, believe that I shall succeed.

Although he insisted that he had to forbid his brother, still living in Rio de Janeiro, several times from undertaking the *wout Miami* because the experience, he said, was "unfit" for human beings, Pipo had opted to turn to his Rastafarian beliefs whose positivity would accompany each stage of the *wout* and provide him with the courage and strength to face even the most difficult obstacles. This decision earned him a position of leadership throughout the *wout*. For example, he found himself the head of a group that had to negotiate dangerous waters on makeshift boats *(la lancha)* and traverse mountains to reach Panama.

Indeed, after a few days' journey by bus from Quito, Pipo had set out, like thousands of other people, on the road leading to Medellín to obtain a thirty-day pass from the Colombian migration authorities, allowing him to continue his journey in search of the "American dream" legally. Arriving in Turbo, a town situated on the Caribbean coast in the north of Colombia, on the shores of the Gulf of Urabá, the departure point of *la lancha* to reach Capurganá, Pipo met a woman whose face he recognized: she had been the owner of the house that his mother used to rent in the center of the Gonaïves, a town in the north of Haiti where Pipo was originally from. This unexpected encounter transported him momentarily back to Haiti. As his memories flooded back, Pipo told me how this woman had helped him, his mother, and his young brother when his father left Haiti for the United States to *canter*,[8] a story revealing the deep bonds of solidarity between the two families.

During their reunion, they swapped news about those who still remained in Haiti, those who had died, and those who had taken the *wout* to Brazil, Chile, or the United States, each one trying their luck in search of a *lavi miyò*. But very quickly, their conversation turned to the details of the organization of the *wout Miami* and the strategies that should be employed to cross the Gulf of Urabá and traverse the mountains, rivers, and dense vegetation to reach Panama, strategies that are primarily linked to the composition and choice of group. Often, Pipo told me, joining a group with children or women for a sea crossing or a crossing on foot under very difficult conditions was risky because they slowed progress and complained far more. The choice and composition of the group were, however, based on the need for women to be present to cook for the rest of the group, which should essentially be composed of strong people capable of enduring the extremely harsh conditions of the voyage.

Reaching Panama from Turbo was known as the deadliest section of the *wout Miami*, passing through the green hell of Darién, a region controlled by paramilitary groups and drug traffickers. Videos and photos provoked intense discussions on the Facebook and WhatsApp groups to which I belonged, depicting and testifying to the extreme conditions involved in the crossing, showing a dense rainforest inhabited by wildlife and vegetation unfamiliar to Haitians. This frontier, situated in a hostile tropical environment called the Darién Gap, is extremely difficult to cross, with no road enabling passage from one country to the other—it is also the only missing link on the Pan-American Highway. From Turbo, most migrants could cross the Panamanian border clandestinely either by land, trekking through the Darién jungle for days, or by sea, taking a commercial boat to cross the Gulf of Urubá. According to some stories collected on social networks, many were unable to survive the extreme conditions and the risks of such a treacherous trek undertaken by inexperienced and poorly equipped hikers where serious dangers abound (criminal groups, wild animals, diseases, starvation, or dehydration). The weakest died, abandoned on the path to the cries and tears of their relatives who were powerless to help and forced to continue their journey.

However, this encounter undoubtedly marked the decision of Pipo and the woman to continue the road together, as though "God had sent this person to find me" *(se Bondye ki voye moun sa bay mwen)*, Pipo said to me. While the woman saw in Pipo a young, resourceful man with a positive spirit whom she trusted because she had known him as a child, Pipo saw in her a strong and courageous woman *(fanm vayan)* who could take care of him, cook for him, help him, or nurse him when necessary. A group of a dozen people was formed, and the woman took charge of buying useful accessories, as few and as simple as possible in order not to overload the rucksacks but enough for everyone's survival needs.

Pipo and his travel companions accepted being at the head of the group. They planned the boat trip across the Gulf of Urabá, which would cost $200 per person, a dangerous journey that would last around eight hours and oblige them to complete the *wout* on foot to reach Panama. Pipo smiled as he explained to me that boarding *la lancha* had taken as long as the crossing, which had been two and a half hours because they had to weigh the baggage, each kilo overweight had to be paid, and everyone was advised to wrap their things in trash bags to protect them. However, once at sea, the motor of *la lancha* stopped and then started again after a few minutes as though nothing had happened—*la lancha* had three motors that operated one after the other to save fuel. The sea journey included an overnight stopover on an island to refuel and rest. The slightest swell made the boat lurch on a relatively calm sea, pounding violently, pitching like a walnut shell flush to the water's surface and throwing around the passengers, who screamed in fear, many of them unable to swim, and most of them seasick.

After arriving at Capurganá, the group led by Pipo set out again on foot to cross the mountains and jungles, a story that he told me with a lump in his throat, revealing that *chache lavi* can also become mixed with the meanings and practices of the *mistik* (magic, mystic) when the *wout* requires confronting extreme travel conditions. Indeed, to reach Panama, the group had to trek for five days in the mountains, traversing rivers in flood, wading through mud, sleeping with one eye open amid the jungle, and watching out for scorpions, vipers, or other poisonous snakes. In Capurganá, Pipo had negotiated for an indigenous Colombian guide for the journey at $250 per person. He told me that the small Colombian town contained a vibrant market for smuggling undocumented migrants: "It's a market, they know that you're a foreigner, and they approach you telling you their price. Then you can negotiate. It's a full-blown business in the region." Sometimes, it was necessary to wait days or even weeks because it was more difficult to negotiate with smugglers who refused to enter the mountains on foot due to the military operations undertaken by the Panama government to combat drug trafficking, making the Darién Gap region even more vulnerable. Then, people would say, *la selva come* (the jungle eats), meaning that the jungle can eat those who venture into it due to heavy rain, military operations, or the presence of paramilitary groups or drug traffickers. Once the negotiation was done, the guide asked Pipo to organize his group. The important thing was to have water and to wear boots for walking in case of rain. They set off walking behind the guide, who carried a machete to cut through the foliage and clear a path for the rest of the group. Pipo's account told of a trek on which it was difficult to breathe due to the heat and humidity, even though the guide had planned a break every thirty minutes for them to rest. It was important for the Colombian smugglers to know the group's composition; they would have to guide as far as the Panamanian frontier since the reputation of their business could be jeopardized. Sometimes, Pipo laughed:

> When we took a break to catch our breath and rest, the Indian would ask us to take photos or make videos with our mobile phones to show our families that we were fine, that they were treating us well. But it was primarily to send messages to those who were still behind, saying that so-and-so was a good guy, a good contact, and that once they arrived in Capurganá they should ask for him and make a deal to cross the border.

But Pipo confided to me that this *raketè*[9] abandoned them after a few hours of trekking—without doubt, he said, because the group was too slow owing to the presence of young children—and the group found itself alone, lost, without a map or compass, with no idea which direction to take. As panic and fear set in, the children cried without understanding what was happening; the women screamed, Bible in hand, imploring *Bondye* (God) to save them from this dire situation and not abandon them. But Pipo told me that as the group's leader, he could not let emotions run wild and allow the group to disband, meaning it was up to him to find a solution for them all to survive and reach Panama. With Etienne, they then quickly separated into two groups, marching in parallel but at a shouting distance from each other.

Before resuming the trek, Pipo explained to me that he had spent some time talking to the woman, asking her to persuade the rest of the group to once again back his decision, confiding to her that he had heard a voice, a *ti zwazo* (little bird), singing to tell him the path to follow. It was then that Pipo told me that the woman was a *manbo* (a Vodou priestess) and therefore understood, even with so few things at hand, how to prepare a ritual request for protection from the ancestors, the *lwa* who would accompany them along the journey until they arrived safely. Pipo related that she gathered up some plants *(fèy)* and boiled them on a fire made with three stones *(wòch)* at the foot of a tree *(pyebwa)* that she had carefully chosen. It was not a question simply of *fè maji* (making magic) by combining the important elements of Vodou like *fèy, wòch,* or *pyebwa*, but rather of approaching the *lwa* and asking them for protection to *ouvè wout la* (open the road/route) and help them overcome the obstacles that presented themselves. In this sense, *maji* and *mistik* do not refer to magic in the strict sense, as defined by Richman (2005, 151).

For five days, the song of *ti zwazo* guided Pipo and the group as far as Panama. In this country, some spent three weeks while others had to stay three months, depending on the government's position on managing these irregular migrants. Hearing the song of a *ti zwazo* or practicing a ritual could be a request for protection from the spirits or ancestors to confront the impasses of the *wout.* While reciting Evangelical prayers from a Bible that someone had slipped into their backpack as an object guaranteeing their protection, it was to plead for *Bondye* to save them. *Pran wout la* was thus accompanied by elements of a collective dimension at the spiritual and transcendental levels where the rituals and objects were necessary for the safety of individuals throughout their journey, some even practicing this ritual before their departure (Richman 2005, 152).

Reaching Panama was not easy, and when the government declaration of May 9, 2016, banned the entry of Haitian and Cuban nationals coming from Colombia, it further complicated the *wout* of thousands of individuals who wanted to cross the country en route to the United States. A new impasse had opened on the *wout Miami*, leading to the invention of new strategies to reach the border at more significant expense and inflated negotiating prices.

When Pipo and Etienne arrived in Panama in the small and impoverished village of Yaviza, they were welcomed by the indigenous community, who had little to offer them. They negotiated to cross the border with police officers for $100 each. Pipo and Etienne then took a bus to Paso Canoas, where they split up on the frontier with Costa Rica. Pipo had asked his friend to continue his journey and gave him the rest of his savings ($40) because Etienne, engaged to a young woman who lived in Port-au-Prince, had only her to finance his *wout*. Unlike Pipo, Etienne had no close family in the United States who could help him financially. Pipo knew how difficult it was for a middle-class person living in Haiti to send money to a family member or close friend who finds themselves in *Miami*. Pipo waited a few days while his father sent him $500 by international transfer, an amount that enabled him to negotiate the services of a *raketè* to reach the Nicaraguan border at Peñas Blancas, a journey known by the name of *la ruta del tráfico* (the trafficking route) due to the circulation of drugs, illegal merchandise, and even trafficked humans. Located near the Pacific coast, permitting entry into Nicaragua from Costa Rica via the narrow strip of land, Peñas Blancas is the sole point of passage by land between two countries whose diplomatic impasses often force people to demand their rights through the local media. The waiting time can last between three to four months, while others opt for one of the various clandestine means of crossing the border: on foot over the mountains (for a sum of $1,000–$1,200),[10] by truck or bus (for a sum of $1,500–$1,800), or by boat via Los Chiles (for a sum of $900), often putting their lives at risk.

While some gave up their jobs and used money from redundancy payments or their unemployment benefits to purchase plane tickets from Rio de Janeiro to Rio Branco and the *wout* from the Peruvian border to Quito, others bought their plane tickets using a bank loan or borrowed from friends living in Brazil. However, embarking on the *wout*, in most cases, demands a certain assurance that the family living in the diaspora is ready to fund each stage where the money transfer market has taken off, especially in Quito (Ecuador), Peñas Blancas (Costa Rica), and Tijuana (Mexico).

Thus, the clandestine demand-driven *rakèt* market drove the negotiations between the *raketè* and the people on the *wout*, who appealed to their family members living in the diaspora for assistance to gather the required money for papers or to pay to cross the border from one country to another. The more people wished to cross a border, such as was the case in Peñas Blancas between May and September 2016, the higher the price demanded by the *raketè*, in this instance increasing to $1,800 per person. It also varies according to the means of transport used, whether by boat, private vehicle (car, taxi, truck, and so on), public transport (bus), or by foot along winding roads.

In general, the *wout Miami* lasts between three and six months and costs around $7,500 per person on average, according to the information

I obtained from lengthy conversations on the WhatsApp groups, which strongly featured in everyday life of Haitians in Brazil during the first half of 2016. Equally, I heard similar stories from personal testimonies elaborated to me during my stays in Miami and Haiti. The duration of the journey and the cost of funding it—evoked by individuals through the expression *fanmi-m ap ede-m* (my family is going to help me)—vary according to the intensity of the flows over time. Hence, the stronger the demand, the higher the price asked for by the local *raketè* and the shorter the length of the journey.

While Pipo cited Peru as an example of *vye peyi, peyi lèd*, meaning that the country is ugly, unattractive, and a difficult place to live in, even in transit, he defined Honduras and Guatemala as countries where it was easy to transit illegally for around $150 despite recollections of the violence he witnessed. Although his story narrated with an air of resignation *(rezinye)* the attempts of the local border authorities or the *raketè* to negotiate the price for a pass from Perú to Guatemala, once he arrived on Mexican soil, a certain relief took over as a new *wout* loomed, this time of an administrative kind, organized between the Mexican and US immigration authorities.

Indeed, once he was registered as a Haitian national by the Mexican Federal Police services and obtained a twenty-day pass, Pipo undertook a new trip of a few days by bus to arrive at the Tijuana–San Diego border, where local humanitarian services awaited. These services are deployed at the request of the US Immigration and Customs Enforcement and the US Customs and Border Protection to receive and shelter migrants while they wait for their asylum and refuge decisions at immigration court hearings with black tracking devices on their ankles.

At the base of the wire mesh fence that marks the border between Tijuana and San Diego, many were hesitant to cross the border when the US government declared, on September 22, 2016, the resumption of deportations for Haitians living irregularly in the United States (Charles 2016). The messages sent by US authorities to these migrants and asylum seekers, whom they allowed to enter while promising to expel them, plunged them into uncertainty. On the verge of reaching the end of their *wout*, an epic crossing of half the American continent, from Brazil to Baja California, they were left standing at the San Ysidro border post on the US side, where they applied to live in the United States at high risk of being sent back to Haiti. Some were discouraged from crossing the border and saw their hopes turn to dust as they decided to remain in Tijuana, at the foot of the fence, trapped between the sea and the desert. They had never imagined staying in Mexico, far from the American dream, the dream of becoming a diaspora.

Like most of those on the *wout Miami* received by US immigration officials in San Diego, Pipo spent four months in a detention center before being released in November 2016. Once out, he joined his father in Margate, Florida, with whom he has lived ever since. He waited close to six months before obtaining his work permit (Employment Authorization Document), a period of time he used to take an English course. During this period, he attended three court interviews. He was asked to provide proof of his asylum request or his request for a family reunion, meetings where he was represented by a lawyer hired by his father for $500. In January 2018, the courts granted him two years of temporary US residence based on a family reunion.

Ultimately, being in the diaspora does not entail just sending money or objects to those still living in Haiti, realizing a social project, or organizing the journey of a family member. It also means publicly demonstrating the fact on social media, revealing the behavioral, moral, social, and symbolic value of the success of the *wout.* Many characteristics definitive of the diaspora person and the *gwo dyaspora* (grand diaspora) in the Haitian imaginary are only possible when living in *gwo peyi* (economically wealthy countries) like the United States. *Chache lavi* is only truly possible, therefore, outside the territorial borders of Haiti, making mobility a resource to be cultivated to achieve the individual's social, economic, and cultural progress.

Conclusion

Based on an ethnography of the experiences of mobility on the *wout Miami*, my aim here has been to detail the different strategies used by Haitian nationals to cross borders increasingly regulated by the different nation-states of the Americas. The *wout* may be long, fragmented, dangerous, uncompleted, and redefined throughout the process, and the places of destination may even be rethought. The mobility process is then interrupted by regimes of immobility that prevent migrants from moving in the desired directions, effectively creating involuntary immobility (Carling 2001) that reveals itself as the reality for many people who venture on the *wout.* This impossibility of continuing a journey added to the shame of returning to the country of origin without having succeeded in becoming *dyaspora* has led to a reconfiguration of border towns and the dynamic of internal migrations on the American continent. Thus, as is common in other parts of the world, including Latin America,[11] the actions of intermediaries have become fundamental agents of mobility.[12]

Once arriving from the *wout* to the United States, some search for their first job to start sending money to Haiti. Others work for a few months and try to save enough money to finance a new *wout*, which may take them to Canada, while many are deported to Haiti. Indeed, deportation (or its threat) diverts the *wout* of many people, some of them preferring to wait on the Mexican border at Tijuana, while others (in prison or released on bail) wait for court rulings on their applications for asylum or refugee status. Others are provisionally released on US soil. Such is the fear of deportation that even before obtaining their court rulings, some people undertake the *wout* to the north of New York State to enter Canada, a border country offering more favorable immigration policies and where the risks of deportation are lower.

Furthermore, *pran wout la* shows how the strategies for crossing frontiers—physical, symbolic, or technological—almost always involve resorting to the *raketè*, a constitutive element of Haitian mobility, deeply rooted in their networks and cultural practices. These intermediaries may refer, for some, to a criminal, sometimes even violent, enterprise—likened to those of coyotes—that facilitates the organization of an illegal journey, thus fitting into the discourse on people trafficking and smuggling. The ambiguities surrounding the forms potentially taken by the *raketè* become meaningful as they also represent an alternative to the increasingly rigid legal procedures and an opportunity to access and open new *wout*. Moreover, this criminalist approach has hindered a more acute perception of the complexity of the *raketè* within the dynamics of (im)mobility. In certain contexts, for example, these intermediaries allow the mobility of people, the crossing of a border, the obtaining of a document, plane tickets, information, whether they act as individuals *(kontak, m konn moun, m gen moun)* or as institutions *(ajans or ajans vwayaj)* (Montinard 2019, 205–10). They are, therefore, terms crucial to a better comprehension of the dynamic of Haitian mobility, whether from Haiti or abroad.

The ambiguities surrounding the figure of the intermediary, in particular the *raketè*, are revealed in the complex relations that individuals and families may have with them, sometimes being a friend, a neighbor, a family member, or someone less close recommended by someone else, a representative of a public authority or from civil society, among others. This multiplicity of relations between intermediaries and migrants raises issues of loyalty, obligation, reputation, and secrecy beyond a simple market relation in which people pay for a service.

Finally, the paper proposed to describe and analyze the dynamics of mobility articulated in networks through the stories of people who were leaving or had left Haiti searching for a better life. The mobility of those who leave can contribute to the immobility of those who stay and vice versa, especially when those who embark on the *wout* participate by sending money for the maintenance of those remaining behind or when those who stay behind or who live abroad participate in financing those who decide to embark on a new *wout*. Exploring the *wout* on which people set out means talking about the search for a full life, about the individual and collective mechanisms and strategies that are reinvented and developed each time within networks and spaces in which creativity, hopes, and uncertainties coexist, sometimes generating strong tensions and frustrations *(fristrasyon)*. These strategies and mechanisms define mobility dynamics, forcing us to (re)think of territories, relations, and people.

A hierarchy thus exists between the different *wout*, physical or symbolic, an economic and geopolitical hierarchy, both for people living in Haiti and for those Haitians living abroad. A *wout* may assume different dimensions and be continuously modified from one individual to another, making it a mobile category—not static, associated with territory in a dichotomous manner as though a clear separation existed between the places of origin and the places of destination. The return to a known or original territory is the story of many Haitians who have taken *wout* filled with obstacles, some ending up deported while others have become *dyaspora*. *Pran wout la* is indeed a becoming, a state of construction, a mode of being in movement, which can assume derivations constitutive of mobility (like *ouvè wout la, kite wout la*, among others) as a way of being in different spaces and moments of mobility (Biehl and Locke 2017, 6).

Chache lavi means knowing how to live well *(byen viv)* in the new country of residence and can thus take the form of a search for a better life, a quest to realize one's dream of becoming *dyaspora* one day, where individual and collective strategies are constantly reinvented and rethought around new *wout*.

Notes

1 This research formed part of my PhD in social anthropology at the Museu Nacional of the Federal University of Rio de Janeiro (PPGAS/MN/UFRJ). Furthermore, parts of this article have been published in an earlier work: Mélanie Montinard. 2020. "PRAN WOUT LA: Expériences et dynamiques de la mobilité haïtienne." *Vibrant: Virtual Brazilian Anthropology* 17. https://doi.org/10.1590/1809-43412020v17d503.

2 Estimates of the number of Haitians living in Brazil vary. According to data from the Ministry of Justice in 2019, there are about 107,000 Haitians in the country. Other data from academic studies refer to numbers of more than 180,000 Haitians.

3 Haiti was the first Black republic in the world to be born from the revolt of the slaves. Haiti is today considered the poorest country in the Americas.

4 It should be noted here that the verb *migrate* is entirely absent from the Creole vocabulary, though it exists in French, English, Portuguese, and Spanish, for example.

5 It is estimated that the Haitian diaspora includes more than 1.2 million people in regular movement, or 11 percent of the population, according to the OECD/INURED report published in 2017. However, these figures are underestimated due to the importance of irregular immigration: some estimate that over 2 million Haitians live in the United States undocumented (International Crisis Group 2007).

6 Although polysemic, the word refers to a space of sociability and belonging in which people meet. It may be associated with a gang, a group of people who meet to chat, a group of friends, a musical group, an association, and so on.

7 Brazil responded to the "problem" of the flows of Haitian migrants through government decisions like Normative Resolution n. 97 of 12/01/2012 of the Ministry of Labor and Employment (*Diário Oficial da União*, Section 1-13/01/2012, p. 19), which "provides for the granting of a permanent visa as stipulated under Article 16 of Law n. 6.815, of August 19, 1980, for Haitian nationals" or the interministerial decree of 12/11/2015 (section 1, 48) granting residence to nearly 44,000 people. Also see Vieira (2014, 2017).

8 *Li te pran dlo*, Pipo would say to me, an expression meaning that a person has ventured on a *wout* by sea.

9 The *raketè* occupy a predominant place in the tales of mobility, less because of their potential status as "illegal agents" and more because of their role as actors and facilitators of the mobility of people on the *wout* who may face legal or material restrictions, thereby leading to situations of immobility.

10 A three-day trip if the smuggler carries out his mission honestly since many abandon their groups halfway, forcing the members to turn back and pay for the services of another *raketè*.

11 See the documentary "Casa en Tierra Ajena" (2017), which tells the life stories and dreams of Central Americans forced to emigrate to North America. Available at: https://www.youtube.com/watch?v=AkrZIumTRjI&fbclid=IwAR377JOnVTqlIraDeT4yhbGgToljPplXwc5cbk7-E9wxv9ZlOwXhX3iMlHY.

12 Intermediaries are a central figure in the classic anthropological literature on migration and mobility, in particular with regard to studies of political and economic relations at the local level, and are generally defined as "brokers." See the authors from the Manchester School (Gluckman 1949; Fallers 1955), as well as Eric Wolf (1956) and Clifford Geertz (1960).

References

Anglade, Georges. 1982. *Espace et liberté en Haïti*. Montréal: ERCE and CRC.

Biehl, João, and Peter Locke. 2017. *Unfinished: The Anthropology of Becoming*. Durham, NC: Duke University Press.

Bourdieu, Pierre. 1991. "Introduction à la socioanalyse." *Actes de la recherche en sciences sociales* 90: 3–5.

———. 2003. "L'objectivation participante." *Actes de la recherche en sciences sociales* 150 (5): 43–58.

Carling, Jørgen. 2001. "Aspiration and Ability in International Migration: Cape Verdean Experiences of Mobility and Immobility." PhD diss., University of Oslo.

Cavalcanti, Leonardo, Antonio Tadeu Riberio de Oliveira, and Dina Araújo. 2016.

A inserção dos imigrantes no mercado de trabalho brasileiro. Relatório Anual 2016. Observatório das Migrações Internacionais; Ministério do Trabalho/ Conselho Nacional de Imigração e Coordenação Geral de Imigração. Brasília, DF: OBMigra. https://portaldeimigracao.mj.gov.br/images/dados_anuais/RelatorioCompleto_v8_0512_pagespelhada_comcapa.pdf.

Charles, Jacqueline. 2016. "US Shifts Haiti Deportation Policy and Gives a Warning." *Miami Herald*, September 16, 2016. http://www.miamiherald.com/news/nation-world/world/americas/haiti/article103373227.html.

Diário Oficial da União (DOU). *Seção 1*. November 12, 2015. Last accessed on September 13, 2021. https://www.jusbrasil.com.br/diarios/104076812/dou-secao-1-12-11-2015-pg-48.

Dias, Guilherme M., and Rosa Vieira. 2019. "Os limites da apropriação de um léxico migratório internacional no Brasil." *Dilemas-Revista de Estudos de Conflito e Controle Social* 3 (Edição Especial): 151–72.

Fallers, Lloyd A. 1955. "The Predicament of the Modern African Chief: An Instance from Uganda." *American Anthropologist* 57 (2): 290–305.

Foucault, Michel. 2004. *Sécurité, Territoire, Population*. Paris: Seuil.

Geertz, Clifford. 1960. "The Javanese Kijaji: The Changing Role of a Cultural Broker." *Comparative Studies in Society and History* 2 (2): 228–49.

Glick-Schiller, Nina. 2011. "Locality, Globality and the Popularization of a Diasporic Consciousness: Learning from the Haitian Case." In *Geographies of the Haitian Diaspora*, edited by Regine O. Jackson, xxi–xxix. New York: Routledge.

Gluckman, Max. 1949. "The Village Headman in British Central Africa." *Africa* 19 (2): 89–106.

International Crisis Group. 2007. "Peacebuilding in Haiti: Including Haitians from Abroad." *Latin America & Caribbean*, Report 24, December 2007. https://www.crisisgroup.org/latin-america-caribbean/haiti/peacebuilding-haiti-including-haitians-abroad.

Joseph, Handerson. 2015a. "Diáspora: as dinâmicas da mobilidade haitiana no Brasil, no Suriname e na Guiana Francesa." PhD diss. Federal University of Rio de Janeiro.

———. 2015b. "Diáspora. Sentidos sociais e mobilidades haitianas." *Horizontes Antropológicos* 21 (43): 51–78. https://doi.org/10.1590/S0104–71832015000100003.

Ministério do Trabalho e Emprego (MTE). 2012. Resolução Normativa n. 97, 12 janvier 2012. Diário Oficial da União, 13/01/2012, Section 1, 19: "dispõe sobre a concessão do visto permanente previsto no art. 16 da Lei no 6.815, de 19 de agosto de 1980, a nacionais do Haiti."

Montinard, Mélanie V. L. 2019. "Pran wout la: dinâmicas da mobilidade e das redes haitianas." PhD diss., Federal University of Rio de Janeiro.

Neiburg, Federico. 2017. "Vidas incertas. Perspectivas etnográficos sobre a economia real." Lecture. *Seminários do DAN*. June 21, 2017. Museu Nacional (UFRJ), Rio de Janeiro.

———. 2019. "Buscando a vida, na economia e na etnografia." Conferência apresentada para promoção a Professor Titular. Departamento de Antropologia do Museu Nacional da Universidade Federal do Rio de Janeiro, 2019. Mimeo.

OECD/INURED. 2017. *Interactions entre politiques publiques, migrations et développement en Haïti. Les voies de développement.* Paris: Éditions OECD. http://dx.doi.org/10.1787/9789264278844-fr.

Richman, Karen E. 2005. *Migration and Voodoo.* Florida: University Press of Florida.

Vieira, Rosa. 2014. "Itinerâncias e governo: a mobilidade haitiana no Brasil." MA thesis. Federal University of Rio de Janeiro.

———. 2017. "O governo da mobilidade haitiana no Brasil." *Mana Estudos de Antropologia Social* 23 (1): 229–54.

Villalobos Vindas, Ivannia (Director). 2017. *Casa en Tierra Ajena. Un Documental Sobre Migración Forzada en América Central.* UNED. https://www.youtube.com/watch?v=AkrZIumTRjI&fbclid=IwAR377JOnVTqlIraDeT4yhbGgToljPplXwc5cbk7-E9wxv9ZlOwXhX3iMlHY.

Wolf, Eric R. 1956. "Aspects of Group Relations in a Complex Society: Mexico." *American Anthropologist* 58 (6): 1065–78.

Zolberg, Aristide, Astri Suhrke, and Sergio Aguayo. 1989. *Escape from Violence: Conflict and the Refugee Crisis in the Developing World.* Oxford: Oxford University Press.

9

Contesting Borders, Protesting Deportation: The Refugee from Darfur in Amman and Cairo

Elena Habersky

"We are waiting for a decision."

December 14, 2015: I accept the social media friend request of a former English student, Adam,[1] who thanks me for taking the time to talk to him. I wish him health and good things; I know where he is at. "Yes, my dear, I am doing well. Just, the problem is that nowadays we have been protesting in front of the UNHCR [United Nations High Commissioner for Refugees] in Jordan." I ask him what the goal of the protest is and if it is the same information being spread on social media and small national news sites. "The main issue of the Sudanese is resettlement and to be looked after like the other refugees in Amman . . . We want to show the world that there are [still] Sudanese refugees, and we receive nothing from UNHCR while the other refugees [Syrians, Iraqis] are receiving money to live. We want people to see our problems, that they are the same as others. I hope there can be equality between us."

The next time I talk to Adam is December 17, 2015. "Hi dear, so yes, right now we are near the airport, maybe there is one kilometer between us, and now we are still waiting for a new decision." The decision Adam and his compatriots are waiting for is whether they are going to be deported to Sudan. The decision taken to ultimately deport this group of individuals throws two groups into disarray: those Darfuris who remain in Amman and those who are deported to Khartoum, eventually making their way to Cairo via irregular means.

Introduction

In Fall 2013, I met a close-knit group of people seeking refuge from Darfur in the low-income east Amman neighborhood of Ashrafiya. Three times a week, migrant communities would descend on a local school for three-hour English classes. These students were bright and intuitive, wanting to deeply understand the nuances of English words such as human rights, genocide, and reparations—no easy task for any teacher. What would transpire would be almost a decade-long friendship that would traverse not only the capital city of Amman but that of Cairo.

Through the desire to escape a war-torn homeland, many of my former students would experience a protracted and fragmented journey (Collyer 2010), taking them from Darfur to Amman, and for a large segment, back to Khartoum and onward to Cairo. Throughout their journey, many say they long to return to Darfur, a paradise in their minds, or be resettled to a country in the Global North to start their lives anew with some semblance of normalcy. Even in their countries of refuge in the Global South, international institutions and governments provide no clear safe haven, leaving them stuck in limbo with dreams of another life. In order to survive in this limbo, the fractured communities organize themselves in a specific environment for safety, something which they have not had for many years. This chapter will then look at how fragmented journeys, and ultimately other parts of migration like border regimes and deportation, stem from such journeys. As border regimes and externalization policies are the main causes of such journeys and deportation is one of the constitutive components of the journey at both the individual and systemic levels, my interlocutors found themselves resisting such systems through their protesting but also while living their lives as refugees in an urban setting.

To better examine the fragmented journeys undergone by those fleeing the conflict in Darfur, Michael Collyer has shown in his "Stranded Migrants and the Fragmented Journey" (2010) how such journeys are becoming more common for forced migrants as more groups are making overland journeys. Fragmented journeys are journeys that are not linear, from point A to point B, but rather take place over time, often many years, and are broken down into a number of various stages that may have different motivations and statuses. Part of the reason for fragmented migration comes from the various containment policies practiced by countries in the Global North, including strict immigration controls, which force asylum seekers and refugees either to continue their journeys from places like North Africa or to set out on dangerous journeys through smuggling and trafficking routes.

As refugees caught in the border regime, they inherently give resistance to and extend the national borders of their daily lives. They are not intrinsically tied to Darfur, Cairo, or Amman, as their relationships, aid received, and hopes for the future cannot be tied down to one particular location. Yet, these lives are very much determined by the laws and organizational structures within their places of asylum, which causes them to sometimes win and sometimes lose their fight against the so-called refugee protection regime that often does anything but protect. While those seeking refuge from Darfur in Amman have been included in more local organizations' programming,[2] their deportation from Jordan (Sweis 2015) and the subsequent hands-off approach of the Egyptian government toward refugees (Norman 2020) have made their lives anything but "enjoyable" in their right to seek and enjoy asylum (Edwards 2005).

Dar Fur: The House of Fur

The conflict in Darfur that spurred the "Save Darfur" movement in the Global North and caught the attention of politicians and activists worldwide can be easily traced back to 2003.[3] The Darfuris, who had been suppressed for decades by the ruling elite, were calling for increased infrastructure in the region, a stake in the government, and part of the oil wealth that the government possessed (Flint and De Waal 2008). Perhaps the biggest colonial legacy of the British is the extreme power imbalance between the administrative capital, Khartoum, and the rest of the country, particularly the West Darfur State, leading to the absence of robust infrastructure (Cockett 2010).

The Sudanese government violently suppressed the protests, sending in government forces and armed guerillas to quash any rebellion as well as financially supporting the Janjaweed militia, leading to attempts at genocide. Both main rebel groups in Darfur, the Sudan Liberation Army/Movement (SLA/M) and the Justice and Equality Movement (JEM), have since echoed the calls of the wider protests for equal representation in the Khartoum government and an end to economic disparity (PBS NewsHour 2008). The violence has caused thousands of people to flee Darfur, often to Internally Displaced Peoples (IDP) camps in other parts of Darfur and greater Sudan, as well as to the neighboring country of Chad.

Musa, another close friend, described one cold winter evening in Amman the day his village was attacked in Darfur. He speaks Arabic to me, but only in numbers. He later tells me they are dates and times: the dates and times his village was attacked in Darfur. Musa remembers the time, which was carefully calculated by the militia to attack his village, right before *fajr*, the daily dawn prayer for Muslims. The village remained clothed in darkness while his fellow village men and women were rousing from their sleep to head to the mosque and perform *wudu'*, the ritual cleansing performed before prayer. Then gunfire, followed by mass confusion. People began running in every direction, unsure of what was going on and where to go. After the village was looted, all the huts were set on fire and burned to the ground so that no trace would remain of a once-thriving village. Then, the militia went off to follow the people who had fled on foot. Before long, they were caught. Musa spent sixteen days hiding near the border with Chad. The year was 2003, and it would take him ten years before he was able to come to Jordan in search of a better life. But life, according to him, is difficult.

Those I met from Darfur in Amman felt they had no other option but to flee, not only the violence in Darfur, which is still occurring to this day (Steers 2020; Dahir 2021; Calvin-Smith and Di Biasio 2021), but Sudan entirely if they were to have any safety.

Escape to Amman

Wanting to be anywhere besides Sudan, a country whose government the Darfuris could not trust, many looked for means of escape. While Egypt seemed to be a clear destination choice based on its geographic location and the long-standing history between the two countries, what would become known as the Mustafa Mahmoud Park incident in the Mohandiseen neighborhood of Cairo in 2005 changed the route of many. The incident involved the killing of at least twenty Sudanese individuals.[4] It was sparked by a three-month protest in front of a UNHCR office attempting to draw attention to the racism, poverty, and high unemployment faced by Sudanese refugees in Egypt, by calling for resettlement to a third country (Whitaker 2005). According to my interlocutors, the news from this incident changed the route of many seeking asylum to Jordan instead of Egypt. Others realized they could take advantage of the lax visa policies to enter Jordan for medical tourism, where they could then apply for asylum (Murphy et al. 2016).

A City of Refuge

Amman is a city of refugees. Modern-day Amman is rather young, established as the capital of the Emirate of Transjordan in the early twentieth century. Since its infancy, it has been a place of refuge for people fleeing violence and conflict. Two such groups are the Circassians and Armenians. When Jordan was dominated by the Ottoman Empire in the early nineteenth century, the Ottomans welcomed Circassians fleeing from the Russian Empire who were no longer welcomed by the Christian rulers on account of their Muslim heritage (Shami 1994). Many Armenians also migrated in the early twentieth century. These individuals were welcomed as they were pushed out by the Ottoman Empire during the 1915 genocide (Derderian-Aghajanian 2009). Contemporary migration flows can be traced back to Palestinians fleeing across the Jordan River after the creation of the state of Israel in 1948. Since the late 1980s and early 1990s, waves of other neighboring communities have been assembling new lives in Amman, the foremost being Iraqis who have fled across the neighboring border beginning with the First Gulf War. Subsequent waves came during the Second Gulf War in 2003 and 2006, in the aftermath of the US invasion, and after the takeover of Mosul by the Islamic State of Iraq and the Levant in 2014. Migrating in waves have also been Syrians, some before the Syrian Civil War, and many after 2011 who were automatically registered under the UNHCR as prima facie refugees without having to go through a Refugee Status Determination (RSD) interview in order to access survival assistance or other UNHCR-promoted durable solutions (Lenner and Schmelter 2016).[5]

Many migrant groups gaining refuge in Amman in the last fifteen years have been fleeing protracted conflicts that show no signs of abating. These groups include Yemeni and African migrant communities of Darfuris and Somalis. Unlike the other refugee communities residing in Amman, the migration of Darfuris to Jordan is unusual due to there being no long-standing historical or geographical ties binding the two states together. This is but one of many reasons why life in Amman is extremely challenging for them, casting them as outsiders within the city. Of course, it is important to note that Arab refugees also face the ongoing possibility of ostracization in Amman. However, the Darfuris feel that their blackness and geographic distance from home place them further away from the Arab Jordanians than their fellow Arab refugees.

Anywhere but Here

As many seeking refuge from Darfur have dreams to either return or venture on to a third country of resettlement, many feel stuck in Amman, unable to work regularly or receive services from international and national organizations. Even the 1951 Convention Relating to the Status of Refugees can only provide so much protection, as UNHCR can only work so well in carrying out its mandate in a system where state deliberations are circumscribed by their national sovereignty, as well as legal, economic, and political interests (Federico and Hess 2021). This is especially true in countries like Jordan, which is not a signatory to the Convention, and Egypt, which has made reservations to certain articles. Many refugees in Jordan are at the whim of local organizations for assistance, many of whom are underfunded, while those in Egypt are at the mercy of a state that has, until now, taken a hands-off approach to refugees (Norman 2020).

As mentioned previously, Darfuris in Amman and Cairo commonly aspire to be resettled to a third country in the Global North. However, border externalization policies have become a great obstacle to their life plans, posing harsh restrictions on refugees, often pinning them to a singular location, which leads many refugees to take on perilous and often uncertain journeys in order to avoid such restrictive border regimes. Now living in a type of limbo in which they feel unable to create a stable and secure life, they feel the only way to claim their rights is to resist.

Life in Amman has been extremely challenging for my interview partners. Besides the outward issues of racism and xenophobia, material and financial support is dwindling, and even more so for minority refugees whose conflicts have disappeared from global news reporting. Work is prohibited in the formal sector, so many Darfuris work in the informal labor market, opening themselves up to exploitation and abuse. They feel betrayed, both by UNHCR, who they believed would protect and provide for them, and also by their hosts, who view them with suspicion. However, they also believed that standing up for their own dignity was a worthwhile fight to undertake. Unfortunately, this decision ultimately led them to a greater and more fragmented journey.

The Deportation

On the morning of December 16, 2015, many in Amman awoke to the news that approximately 950 refugees from Darfur had been rounded up in the early hours of the morning from in front of the UNHCR offices in Khalda, an affluent neighborhood in west Amman and loaded onto buses to undergo deportation. It had been thirty days since these individuals staged their peaceful sit-in outside the premises of the UN agency, their colorful patchwork tents of bright oranges and deep blues marking their territory (Malkawi 2015). For many months, even years, for some, Darfuris in Amman believed they were being treated as second-class refugees within the larger refugee community, which was mainly constituted of fellow "brother" Arabs to the Jordanians: Syrians and Iraqis. However, it is contested that these communities are treated in a welcoming manner (Su 2013). Those from Darfur were also protesting the harassment and violence that was oftentimes perpetrated against them due to their racialization in the host community (Su 2014). Throughout all of this, the protesters wanted acknowledgment from the media to show the world their plight. While there were articles written by smaller news agencies during the protest, journalist friends with ties to the community were denied access to the protest by security forces patrolling the campsite (Lucente 2015; Staton 2015).

Their demands, addressed to "those working in the humanitarian field," were written in English to signal to UNHCR, but also to Western aid and embassy workers from countries where they wished to be resettled, that their grievances were directed toward them. One of the young community leaders wrote a document listing their demands, beginning by stating, "We the Sudanese refugees still face discrimination, marginalization, and are being ignored by the UNHCR." Their demands, as listed, were as follows: an urgent housing allowance; faster interviews for refugee status determination, resettlement, and immigration processing; reopening of any closed files through UNHCR; provision of comprehensive healthcare services; and opening of more resettlement opportunities for Sudanese refugees in more than one country.

It is no surprise that the community resorted to protest. It has long been the case that migrants, civil society actors, and other institutions contest and protest both containment policies as well as the deterioration of services and protection mechanisms in the country of asylum (Stock, Üstübici, and Schultz 2019). As has been seen before, though, migrants are oftentimes seen as a burden rather than as part of the fabric of the society, especially when they cause disruption (Castles 2010). Although part of the reason for the protest was a lack of resettlement places, causing immobility, other types of mobility, such as resistance to said externalization policies, are seen as a security threat by the nation-state (Stock, Üstübici, and Schultz 2019). Whereas certain forms of immobility can be fundamentally tied to containment policies and border externalization from the Global North, states in the Global South also make their own autonomous decisions based on their own interests, as Genç, Heck, and Hess (2019) found in Turkey. Faist (2019) also argues that so-called countries of transit are not simply passive actors toward the wishes of the Global North but also have their own room to maneuver and make their own calculated decisions. One can see this in Jordan, which closed its borders to Syria, citing security concerns, though it has been argued that the country was attempting to show the international community the supposed magnitude of the influx in order to receive large sums of financial assistance (Tsourapas 2019).

After one month of peacefully protesting outside UNHCR, contesting the policies that hindered their mobility, as well as the policies that hindered their access to services, hundreds of Darfuris were loaded onto buses with hands visibly zip-tied. Photographs quickly circulated, and news started trickling out that those loaded onto the buses were headed to the airport, where they were to be deported back to Khartoum. Calmness quickly turned into chaos.

The deportation process would not be straightforward and linear. At first, friends and contacts still in Amman received messages from their protesting friends that they were being loaded onto buses in order to be resettled in Canada (Gibreel 2017). They believed a solution to their demands had finally been found. Some of the protesters even called and messaged friends during the early hours of the morning that they should come to the UNHCR office to not miss this opportunity. The soon-to-be deportees quickly became aware, though, that this was not the case. In truth, the protesters were heading to an unknown location, which would lead to their deportation. In total, the deportation process would take two days and an abundance of misinformation. Ariane Rummery, a spokeswoman for the UNHCR in Geneva, stated during the deportation process, "We have appealed and we continue to appeal to the government to stop the deportations from Jordan of Sudanese nationals who are registered with UNHCR as refugees and asylum seekers" (Sweis 2015). In addition, Joe Stork, Deputy Middle East Director for Human Rights Watch, stated that "Jordan should not punish these Sudanese merely because they protested for better conditions and for resettlement considerations" (Sweis 2015).

Some individuals were able to escape during the chaos of the final stop before the airport, a shipping warehouse, where they were able to climb over barbed wire fences and escape in the night back to Amman. For some, the return home to worried friends involved hours of walking with lacerations to their hands. Others, unable to escape, barricaded the doors of the warehouse, leading to confrontations with security forces that resulted in beatings and the use of tear gas. On the morning of December 18, 2015, much to the dismay of friends, colleagues, activists, and concerned individuals in Jordan and across the world, the deportation finally commenced for the remaining individuals in the warehouse, their resistance having, for the time being, been quashed (Sweis 2015).

Arrival in Sudan

On December 20, 2015, I received a message from a friend and former student, Izzeldin. When he messaged me, he said he had safely arrived in

Khartoum, but not everyone on his flight from Amman was as lucky. This was the first sign of life that anyone had heard from any of the individuals who had been deported two days before. Izzeldin told me that he felt hopeless being back in the country he tried so hard to flee but that there was nothing anyone could do, it was all in God's hands.

Izzeldin, like most of the deported, also knew he needed to travel immediately after landing. "I want to travel, but I can't go by the right way. There's no way open." Izzeldin knew if he wanted to move, if he wanted to leave, it would have to be via smugglers across the border with Egypt. "Nothing is safe in Sudan. Me so sad for everything. Am trying to go to South Sudan, it is safer than Sudan." I had heard through his friends in Amman that, eventually, Izzeldin made it safely overland to Egypt. He also talked about his friends who were taking boats to Libya, some of whom were our mutual contacts. "Some of friends are taking [boats], yea, I can't forget. They are my friends." I had already heard the terrible news that some of those who had taken boats had not survived the journey, and again Izzeldin reminded me that there was nothing we could do about these unpredictable journeys. "What can we do. This has come from our God, nothing in our hands." Thankfully, a few did survive the journey and are now safely living in France and England. Others who were deported attempted to go back to Darfur and see family and friends they had not seen for years. Sadly, some were killed in violent encounters with local militias, pointing to the protracted nature of the conflict they had originally fled and the unfortunate outcomes of such long, unpredictable journeys.

Another friend and de facto community leader, Musa, described what had happened to him and a few others once they landed at the airport in Khartoum. "I could not go back to Darfur, there is too much violence still happening, though one of our brothers went back. He was killed, did you see? When we landed at Khartoum Airport, we all went through six-hour interviews. Yea, from 12 to 6, six-hour interviews! The security forces had files on all of us, they knew everything about our backgrounds, including our education levels, so I knew I couldn't lie. They even had what major I studied in college. They asked us many questions. Where we were from in Darfur, where our families were now, if we still had any contacts located in Sudan, what was our address, where we were planning on staying. They took our fingerprints and told us we could not leave Sudan legally for an undefined period. I knew if I would leave, which I needed to, I would have to leave by smuggling to Egypt. I only had five Jordanian dinar in my pocket as they took all my belongings during the deportation, so I had to give the person [the smuggler] my Samsung phone that my cousin sent to me when I was in Amman. It was Cricket Wireless. I really miss that phone."

The Deported Start Anew in Cairo

On the other side of the Red Sea, opposite Jordan, Egypt and Sudan have a long and, oftentimes, complicated history, but also a long-standing one of migration due to historically porous borders between the two countries when the line of demarcation was fluid and changing. Modern-day migratory movement between Egypt and Sudan has a strong, albeit complex fluidity due to the Wadi al-Nil Agreement of 1976 and the Four Freedoms Agreement of 2004 (Jacobsen, Ayoub, and Johnson 2014). While many Sudanese were able to cross into Egypt under the terms of the agreements, such as the waiver of visas and easier access to residency and work permits, Sudanese asylum seekers from various conflicts like those in Darfur and South Sudanese fleeing conflict in their new nation-state have needed to register with UNHCR to gain refugee status and the legal entitlements that come with such status. Currently, relations between the two neighbors are strained due to security concerns (El-Mahdawy 2017). While Sudanese who have lived in Egypt for decades may feel well integrated into Cairene society, many new arrivals feel unwelcomed by the hosts, facing issues of discrimination and harassment, and, in worst-case scenarios, violence (Salih 2020). This is true in various aspects of their daily lives, even with all the similarities present between the two countries. Once again, their differences, rather than similarities, cast them as strangers within the urban space in which they have found refuge.

For those Darfuris who entered Egypt irregularly through smuggling channels, life was proving to be just as difficult as in Amman. Musa messaged me during the fall of 2016 that some of his friends were arrested and forcibly returned to Sudan, and UNHCR was moving slowly in getting people out of detention. "UNHCR here like UNHCR in Jordan, just words on paper. Forcible return here and there. Now the police are arresting a lot of refugees. For that some of them today came out and sit in front of UNHCR. Yeah. . . I saw some of [them] arrested in front of my eyes. I'm lucky. Police officer said to them, where's your ID, and even though they got ID of UNHCR they were still arrested."

Upon arrival, many of the deported from Amman began finding each other in order to find housing. As in Amman, it is cheaper in Cairo for different refugee communities to live together in apartments to lower the overall cost of rent and to have a sense of community. While Cairo is nowhere near as expensive as Amman, it is still difficult to afford rent when you are unable to legally work. For many of the now "Deported Darfuri," as they named themselves, living in the neighborhood of Ard al-Liwa' in the Giza governorate was the preferred choice. Built as an informal settlement, Ard al-Liwa' is now a distinct neighborhood home to many refugee populations and Egyptians of a lower economic class. For the "Deported Darfuris," it was a good location to start their lives in Cairo, as it offered them cheap accommodation and communities who would be aware of their situation, including other Darfuris, Sudanese, and South Sudanese migrants and refugees.

Those who do not live in the same apartments in Ard al-Liwa' live in close proximity to one another, with other "Deported Darfuris" living in farther-off areas like Ain Shams, Faisal, and Nasr City, which also offer cheap accommodation and access to the informal labor market. Ahmed is in his early thirties and was awaiting a new baby with his wife, whom he married after he came to Cairo in 2016. Always sharply dressed in a suit, he talks about his new neighborhood as follows:

> Ard al-Liwa' is okay, but there are problems sometimes. Two weeks ago [in late 2017] someone was shot and they died. There are problems between rival gangs and sometimes between the refugees and Egyptians. We try to stay out of it because we don't want any trouble, but we always hear when a problem becomes big. Usually the people don't die though, I hope that does not happen again. I only want something that is cheap and safe for my wife and our new baby. I don't want the baby to grow up in violence, I tried so hard to get away from that.

For those now in Cairo, communication with one another, as well as with those left behind in Jordan, remained crucial to circumventing the border and protection regimes causing such fragmented and complicated journeys. As the migration policies of one country violently disrupted these individuals' lives, they ultimately achieved little in spite of their forced dispersal as a group, always reinventing and challenging the regimes meant to contain them (De Genova, Mezzadra, and Pickles 2015). Sitting one evening with my friend in a local coffee shop, I asked him why the "Deported" were such an insular group and if he saw any possibility of reaching out beyond this tight circle.

> We have to support each other and stay with one another. There are over 200 of us that were deported from Amman who are here now in Cairo. Deep down, we don't trust other Darfuris. Of course, we can talk with them and can be friends, but never deep down. We just

don't know. Other Darfuris, they could be criminals or even spies for the Sudanese government. We always have to be careful what we say to someone we don't know. Listen, there are four groups of Sudanese people in Cairo. You have the diplomats, and they are working for the government. These people you cannot trust. You have the tourists who want to come and travel because Egypt is so close to Sudan. You have the workers who come here to make money and then send the money back home to their families, some of them have been here for a while. And then you have us, the refugees. But even amongst the refugees, you can't trust everyone because you just don't know. So we know these four groups exist and that we are a part of these groups, but still we are always careful. And of course, we hang out with other African populations. We go out to cafes, we go to watch football matches together or we play in the fields in Ard al-Liwa'. But in our homes, it's usually just us, there is no big mixture. But, we do this for survival. We have a WhatsApp group called "The Deported" and it has all the people who came here from Amman. We message each other every day. And we always talk to our friends in Amman every day, I miss my friends very much. But people usually stay to themselves and this is okay. It is to keep our identities alive at all costs. If you lose your identity, you are in the middle of nowhere, you are stranded."

Over the few short years they have come to know one another in exile, my interview partners have developed their identities from a different experience, particularly the traumatic deportation experience, which is in contrast with other Sudanese refugees and migrants in Amman and Cairo. This distinct difference, which has also created a hindrance to building a community with others in a trusting way, comes from these experiences during the fragmented journey. With good reason, the "Deported Darfuris" maintain strong communities with those in Amman and Cairo who went through the same experience, making their true location from where they contest neither in Amman, Darfur, nor Cairo.

Conclusion

The story of the "Deported Darfuris," from Sudan to Jordan and deported back to Sudan and onward to Egypt via smuggling, shows that being in exile is more complicated than conceptualizing an imagined future. While many claim they want to return to Darfur, no one has voluntarily attempted to return since they left. While life is difficult in both Amman and Cairo, especially for those seeking refuge from African countries, the communities they have formed have proved vital to their survival and overall happiness and, more importantly, to their resistance to the regimes that attempt to contain and suppress them.

Overall, sharing in the deportation process makes them feel closer to one another, both for those left behind in Amman to pick up the pieces and those who are in Cairo and need to start anew. However, this has not changed the fact that those in Cairo and Amman continue to live life in limbo, ultimately desiring to go somewhere else where they can be settled and have residency, access to services, and basic rights. While Darfur may be the imaginary bonding point, the homeland, their fragmented journeys are ultimately not tied to a specific place, which we can see from the WhatsApp groups in use between Amman and Cairo. Contesting the regimes that attempt to pervade their lives is what ultimately connects them to one another.

Notes

1 All names in this essay have been changed by the author to ensure confidentiality.

2 For more information on organizations in Jordan that are including minority refugee groups in their programming, please see: https://www.collateralrepairproject.org/meet-our-community-2/ and https://sawiyan.org/.

3 While it can be easily traced back to 2003, there were groups working on bringing attention to Darfur before the current conflict. For more information on the history of the Save Darfur Coalition, please see: https://www.pbs.org/wnet/worse-than-war/get-involved/organizations-save-darfur-coalition/77/.

4 These numbers have been disputed by Sudanese community members, with some leaders claiming that over 100 Sudanese lost their lives in this incident.

5 Even though Syrians do not have to go through the RSD process, this does not mean that their access to services or resettlement is high. Only those deemed "most vulnerable" get resettled, and, even then, we know that less than one percent of all refugees get resettled on an annual basis.

References

Calvin-Smith, Georja, and Laura Di Biasio. 2021. "Sudan: Death Toll from West Darfur Clashes Climbs to 56." France 24, April 6, 2021. Accessed May 18, 2021. https://www.france24.com/en/tv-shows/eye-on-africa/20210406-death-toll-in-western-darfur-clashes-climbs-to-56-eye-on-africa.

Castles, Stephen. 2010. "Understanding Global Migration: A Social Transformation Perspective." *Journal of Ethnic and Migration Studies* 36 (10): 1565–86.

Cockett, Richard. 2010. *Sudan: Darfur and the Failure of an African State.* New Haven: Yale University Press.

Collyer, Michael. 2010. "Stranded Migrants and the Fragmented Journey." *Journal of Refugee Studies* 23 (3): 273–93. https://doi.org/10.1093/jrs/feq026.

Dahir, Abdi Latif. 2021. "Violence in Sudan's Darfur Region Dims Hopes of a Long-Sought Peace." *The New York Times*, January 19, 2021. https://www.nytimes.com/2021/01/19/world/africa/sudan-darfur-violence.html.

De Genova, Nicholas, Sandro Mezzadra, and John Pickles, eds. 2015. "New Keywords: Borders and Migration." *Cultural Studies* 29(1): 55–87.

Derderian-Aghajanian, Ani. 2009. "Armenians' Dual Identity in Jordan." *International Education Studies* 2 (3).

Edwards, Alice. 2005. "Human Rights, Refugees, and the Right 'To Enjoy' Asylum." *International Journal of Refugee Law* 17 (2): 293–330. https://doi.org/10.1093/ijrl/eei011.

Faist, Thomas. 2019. "Contested Externalisation: Responses to Global Inequalities." *Comparative Migration Studies* 7. https://doi.org/10.1186/s40878–019–0158-y.

Federico, Veronica, and Sabine Hess. 2021. "Protection Regimes: A Critical Analysis." Working Paper: *Respond Horizon*, 2021/76. https://www.academia.edu/45617695/Protection_Regimes_a_Critical_Analysis.

Flint, Julie, and Alexander De Waal. 2008. *Darfur: A New History of a Long War*. Revised & updated. New York and London: Zed Books.

Genç, Firat, Gerda Heck, and Sabine Hess. 2019. "The Multilayered Migration Regime in Turkey: Contested Regionalization, Deceleration and Legal Precarization." *Journal of Borderlands Studies* 34 (4): 489–508. doi: https://doi.org/10.1080/08865655.2017.1344562.

Gibreel, Dana. 2017. "After the Deportation: How Jordan Left Sudanese Refugees to Death and Separation." *7iber* | حبر. September 19, 2017. Accessed March 20, 2018. https://www.7iber.com/politics-economics/after-the-deportation/.

"In the Name of Sudanese Refugee Community in Jordan." 2015. Document, in the author's possession. December 2015.

Jacobsen, Karen, Maysa Ayoub, and Alice Johnson. 2014. "Sudanese Refugees in Cairo: Remittances and Livelihoods." *Journal of Refugee Studies* 27 (1): 145–59.

Lenner, Katharina, and Susanne Schmelter. 2016. "Syrian Refugees in Jordan and Lebanon: Between Refuge and Ongoing Deprivation? Dossier: Mobility and Refugee Crisis in the Mediterranean." *Mediterranean Yearbook*. https://www.iemed.org/wp-content/uploads/2021/01/Syrian-Refugees-in-Jordan-and-Lebanon-between-Refuge-and-Ongoing-Deprivation.pdf.

Lucente, Adam. 2015. "Is UNHCR in Jordan Discriminating against Sudanese Refugees?" *Al Monitor*, December 2, 2015. Accessed March 20, 2018. https://www.al-monitor.com/pulse/fr/contents/articles/originals/2015/12/sudan-refugees-jordan-unhcr-discrimination.html.

el-Mahdawy, Hadeer. 2017. "Sudan to Require Visas for Egyptian Men Aged 18–49." *Ahram Online*, April 7, 2017. http://english.ahram.org.eg/NewsContent/1/64/262493/Egypt/Politics-/Sudan-to-require-visas-for-Egyptian-men-aged-.aspx.

Malkawi, Khetam. 2015. "Sudanese Refugees' Protest near UNHCR Enters 17th Day." *Jordan Times*. December 8, 2015. Accessed March 20, 2018. http://www.jordantimes.com/news/local/sudanese-refugees'-protest-near-unhcr-enters-17th-day.

Murphy, Emma, Will Todman, Abbie Taylor, and Rochelle Davis. 2016. "Sudanese and Somali Refugees in Jordan." *Middle East Report* 279. Accessed March 20, 2018. https://merip.org/2016/09/sudanese-and-somali-refugees-in-jordan/.

Norman, Kelsey P. 2020. *Reluctant Reception: Refugees, Migration and Governance in the Middle East and North Africa*. Cambridge: Cambridge University Press.
PBS NewsHour. 2008. "Origins of the Darfur Crisis." PBS: Public Broadcasting Service, July 3, 2008. Accessed November 14, 2020. https://www.pbs.org/newshour/politics/africa-july-dec08-origins_07-03.
Salih, Zeinab Mohamed. 2020. "Dozens of Sudanese Migrants Held in Cairo after Protests." *The Guardian*, November 12, 2020. https://www.theguardian.com/global-development/2020/nov/12/dozens-of-sudanese-migrants-held-in-cairo-after-protests.
Shami, Seteney. 1994. "Displacement, Historical Memory, and Identity: The Circassians in Jordan." *Center for Migration Studies Special Issue* 11: 189–201. https://www.semanticscholar.org/paper/Displacement%2C-Historical-Memory%2C-and-Identity%3A-The-Shami/c74a21d21ce222251416d3f6ecd229725948329.
Staton, Bethan. "Sudanese Refugees in Jordan Pitch Tent City to Protest against 'Discrimination.'" *Middle East Eye*. December 11, 2015. Accessed March 20, 2018. http://www.middleeasteye.net/news/sudanese-refugees-jordan-pitch-tent-city-protest-discrimination-21780425.
Steers, Julia. 2020. "Inside the Forgotten War in Darfur, Where the Killing Never Stopped." VICE News. 13 July 2020. Accessed November 15, 2020. https://www.vice.com/en/article/935db5/inside-the-forgotten-war-in-darfur-where-the-killing-never-stopped.
Stock, Inka, Ayşen Üstübici, and Susanne U. Schultz. 2019. "Externalization at Work: Responses to Migration Policies from the Global South." *Comparative Migration Studies* 7. https://doi.org/10.1186/s40878–019–0157-z.
Su, Alice. 2013. "How Do You Rank Refugees?" *The Atlantic.* November 22, 2013. Accessed March 20, 2018. https://www.theatlantic.com/international/archive/2013/11/how-do-you-rank-refugees/281771/.
———. 2014. "Sudanese Refugees Are Being Attacked by Angry Mobs and Police in Jordan." VICE News. November 21, 2014. Accessed March 20, 2018. https://news.vice.com/article/sudanese-refugees-are-being-attacked-by-angry-mobs-and-police-in-jordan.
Sweis, Rana F. 2015. "Jordan Deports Sudanese Asylum Seekers, Spurring Outcry." *The New York Times*, December 18, 2015. Accessed February 11, 2018. https://www.nytimes.com/2015/12/19/world/middleeast/jordan-deports-sudanese-asylum-seekers-spurring-outcry.html.
Tsourapas, Gerasimos. 2019. "The Syrian Refugee Crisis and Foreign Policy Decision-Making in Jordan, Lebanon, and Turkey." *Journal of Global Security Studies* 4 (4): 464–81.
Whitaker, Brian. 2005. "20 Killed as Egyptian Police Evict Sudanese Protesters." *The Guardian.* December 31, 2005. Accessed March 20, 2018. https://www.theguardian.com/world/2005/dec/31/sudan.brianwhitaker.

10

"The Gospel Doesn't Know Borders, Neither Do We": Congolese Migration, Religion, and Entrepreneurship in Three Metropolises: Istanbul, Rio de Janeiro, and Guangzhou

Gerda Heck

Introduction

It's Sunday afternoon in a hotel conference room in Istanbul.[1] Pastor Christine is preaching in front of approximately eighty churchgoers, most of them Congolese. Pastor Christine is a London-based Congolese trader who makes use of her frequent business trips to Istanbul to preach from time to time in the Église Charité, a Congolese revival parish in Istanbul. A few years ago, her husband had transited Istanbul and the parish on his way to Europe. On a visit, she reassures the churchgoers during one of her sermons, "So I am aware of the hardships you are facing here in Istanbul . . ." Cesar, a parishioner who is translating for me from Lingala into French during the worship, explains, "She is one of my clients." Then he excuses himself and says he must leave for the airport to pick up one of his trading clients.

This is a scene I witnessed during one of my multiple visits to the Sunday worship of Église Charité (in English: Charity Church), the Congolese Pentecostal church based in Istanbul, between 2011 and 2016. "With some of our parishioners, we have managed to establish a small business enterprise to make our living here in Istanbul," Pastor Alain, the head of the parish, explains to me later:

> That means, we take care of traders who come to Istanbul, we pick them up at the airport and put them up in a hotel, we accompany them to the sellers, we bargain and translate for them. And of course,

a lot of our clients are believers and some of them are pastors, so we all meet in the church on Sunday.

Pastor Alain arrived in Istanbul in 2003 after having lived in Lebanon for some time and traveling to Syria. In Istanbul, he initially applied for asylum at the United Nations High Commissioner for Refugees (UNHCR). Like many Congolese, he was hoping to get resettled after a positive decision. During our first meeting in 2011, he was still waiting for a decision. In 2006, he founded his French- and Lingala-speaking[2] church in Istanbul, Église Charité. In the same year, he launched a small informal business enterprise in which they organized and intermediated for predominantly Congolese traders but also other French-speaking African traders who regularly travel to Istanbul. "Our main task is praising God, preaching the gospel, and spiritual assistance," Pastor Alain tells me in the interview, "but, furthermore, we try to enlarge the parish, connect with other parishes in Turkey and beyond in Africa, Asia, and Europe, and take care of the daily problems of our parishioners. My dream is to someday have my own church building here in the center of Istanbul. But this is still an ambitious enterprise." Pastor Alain had already been waiting a few years to be resettled and was hoping to continue his journey soon. However, at the time of the interview, he was still highly engaged in the Christian religious landscape in Istanbul while simultaneously enlarging religious and economic networks internationally.

A closer analysis here discloses the entanglement of transnational mobility, immobility, personal aspirations, religion, and trade. Drawing on findings from a multi-sited ethnography (Marcus 1995) on the role of new Pentecostal churches on the migration routes of Congolese migrants, which I have carried out since 2010,[3] I will delve into the complex interplay of religion, global migration, and the urban space of Congolese migrants in three metropolises within the Global South: Istanbul, Rio de Janeiro, and Guangzhou. Doing this, I will describe and analyze how Congolese migrants establish (religious) networks and how religiously inspired aspirations can facilitate transnational migration and "religious place-making" in global metropolises.

In the last two decades, there has been an increasing interest in urban studies of very diverse religious movements in global cities. Thereby, religion is envisioned not only as an agglomeration of religious practices and structures but also as settings in which the religious and the urban are intertwined, mutually productive, and transformative. Similarly, in migration studies, there has been increasing attention to the relationship between religion and migration from the Global South (Anderson 2014). Religious practices play a vital role in migrants' everyday life and community formation (Hüwelmeier and Krause 2010; Garbin 2010; Levitt 2007; Vásquez and Knott 2014), as well as in the creation of transnational religious networks of migrants (Garbin 2010; Levitt 2007). Furthermore, researchers have observed the connection between Pentecostalism and entrepreneurial initiatives and aspirations for new opportunities through migration, in addition to the role of churches in spiritually preparing their congregants for emigration (van Dijk 2012). A significant portion of this research strand still focuses on migration from the Global South to the Global North and is situated in the Northern context, though the larger part of the migration movements is happening within the Global South (Awad and Natarajan 2018). This contribution amends the discussion in the field as it targets migration movements within the Global South.

The scene described at the beginning of the chapter also reveals the crucial role of the church locations for the believers on the migration route. These locations are part of the broader "religious infrastructure" in which believers share religious and day-to-day life experiences and keep up with their religious duties, while the believers are mobile. These locations also serve as reference points. Recent debates on migration infrastructures have emphasized the role of social networks in migration processes by focusing on mediating actors. The mediating actors are situated between the potential migrant and the migration act itself. Xiang and Lindquist (2014) argue that infrastructures of migration have become more and more relevant as migration is more and more conveyed through technologies, institutions, and actors that render and determine mobilities (Xiang and Lindquist 2014, 124). These actors comprise, among others, recruitment agencies, visa brokers, smugglers, humanitarian organizations, and state and non-state actors (Alpes 2017; Cranston, Schapendonk, and Spaan 2018; Lin et al. 2017).

However, the driving force behind the use and the creation of migrant infrastructures is the migrants themselves and their (in)mobilities. In this contribution, I want to take a closer look at the migrants themselves, their aspirations, and the networks they establish. To take the perspectives of the migration as a starting point does not at all mean to neglect the infrastructures or even divorce their acting from social relations. Migrants' mobilities are shaped by social, economic, and political factors at various levels, from family and kinship networks to the local urban context to the national and international levels, but also by their own aspirations and goals.

Framing the concept of "people as infrastructure," the urban researcher AbdouMaliq Simone (2004) expands the concept of infrastructure to people themselves, to an infrastructure that generates possibilities of operating jointly beyond the scope of a particular aim or plan of any individual or group, to people's activities, their unpredictable movements, interactions, connection, and preliminary possibilities (Simone 2004, 407).

Drawing on this, I argue that out of the migratory movement of religious believers, an infrastructure emerges globally, encompassing worship places but also religious articulations, prayers, symbols, music, feelings, meanings, and other linkages. This infrastructure engenders prospects for the believers of acting in common beyond explicit intentions or plans. By doing this, believers make place in the city and participate, (re)create, and transform parts of the urban space in constant negotiations with structures and systems of power. The religious parishes and their urban place-making are closely entangled with the position of Congolese communities in the cities and their emergent daily life and reproduction practices. Thus, when approaching such a fluid network of migrants, it is important to consider the different national, juridical, and economic constraints, as well as the improvisation and impermanence, that characterize their experiences.

In their groundbreaking book *Worlding Cities*, Ananya Roy and Aihwa Ong (2011) put forth a specific idea of "worlding" as an analytical framework of postcolonial urbanism that is capable of examining urban situations, both in their local and global aspects without subjecting them to overly simplistic explanatory models. Their analytical framework breaks with the mainstream "core-periphery model of globalization" (Roy 2009, 824). Accordingly, (religious) "worlding" practices can be seen as interventions within urban as well as national spaces located in everyday practices and understood as a dynamic interplay of "transnational ideas, institutions, actors and practices" (Ong 2011, 4).

Using Ong and Roy's concept of "worlding," as well as Simone's "people as infrastructure," this chapter will expound on how Congolese migrants make use of their religiously motivated aspirations and imaginations to create alternative urban worlds that transcend the city as well as national contexts.

Below, I will briefly outline the emergence of the Pentecostal movement in the Democratic Republic of Congo and trace some of its emerging migration movements. I will then describe the role of religion for Congolese revival Christians in Istanbul, Rio de Janeiro, and Guangzhou and how they navigate and anchor themselves in different national contexts and cities.

Christian Revival in the Democratic Republic of Congo (DRC)

Since the end of the 1980s, simultaneously in other sub-Saharan countries, Pentecostalism has gained a major influence on religious culture in the DRC

(Meyer 2004). Drawing on a "religious awakening," a growing number of churches that are largely called "Église de Réveil"[4] (revival churches) emerged at the end of the Mobutu era against the background of political instability, economic crisis, and escalating violence.

This revival has not only radically altered the religious scene across the country, but it has also become increasingly influential within Congolese public spheres (Pype 2006, 300). Public space and reality, as well as the country's social and moral landscape, have been fundamentally reshaped. Charismatic Christian forms of popular culture, like music, dance, or film, have been introduced. New Christian film and theater productions, as well as radio and TV stations, have arisen, and models of kinship as well as moral agendas of solidarity have been fundamentally transformed (Honwana and de Boeck 2005, 1–2).

Congolese Migration Trajectories—Religious Trajectories

For many decades, Congolese have emigrated toward other African countries, Europe, and North America (de Boeck and Jacquemin 2006; Garbin and Vasquez 2011). Particularly since the late 1980s, more and more Congolese have started to emigrate toward the Global North, Europe, the US, and Canada, though most Congolese emigrants remained on the African continent. Against the background of the collapse of the Mobutu era, political instability, the devastating violence of war, as well as a severe economic crisis, emigration has increased profoundly since then (Schoumaker and Flahaux, 2013). Simultaneously, mobility and migration turned into a major concern globally, especially in the hegemonic states of the Global North. Since the 1990s, in many parts of the world, such as the European Union, the United States of America, Australia, and South Africa, migration policies have become severely restrictive, which has dramatically affected the mobilities of a great part of the world population. Access to global mobility has become one of the most stratifying parameters in today's global society (Bauman 1998). For most Congolese, as with other nationalities, it has become more and more difficult to emigrate on a legal basis to Europe and North America (Heck 2014). In response to these restrictive migration and visa policies in the Global North, as well as new economic opportunities in countries such as Turkey and China, many Congolese started to alter their migration strategies and followed new routes, either undertaking circuitous journeys that involve transiting several countries before reaching their desired destination or targeting new destinations. Some of these journeys take years. Countries such as Turkey or Brazil, for example, which have relatively open visa policies, have turned into new transit and immigration countries during the last two decades. In addition, countries like China have, due to their accessible business visa and trading opportunities, become new destinations for Congolese migrants (Braun 2019).

By virtue of increasing migration, a Congolese religious diasporic landscape has arisen globally. In contrast to their Catholic or Protestant counterparts, revival Christians are able to open up a church without many administrative formalities, official Church permissions, or trained leadership (Kalu 2010, 32). Hence, in the last thirty years, in many metropolises around the world, Congolese revival parishes have been established.

Istanbul—Arriving, Transiting, and Trading

Since the start of the 2000s, Turkey has become a significant immigration center for migrants from Africa, among them Congolese. Although a great part of those arriving regard Turkey only as a transit country, there is a growing number who find themselves stuck as they are unable to continue their journey into Europe (Brewer and Yükseker 2009). Many seek asylum at the UNHCR, hoping to get resettled to a Global North country after their recognition, while others try to journey on to Europe clandestinely. Besides these individuals, Congolese students reside in Istanbul, as well as Congolese suitcase traders who travel back and forth between Istanbul and Kinshasa, but also between Istanbul and Paris, London, Brussels, Johannesburg, or Libreville. Due to its tremendous fluidity, the Congolese community in Istanbul remains small, with the community itself estimating their number between 100 and 150 individuals.

Turkey is a signatory to the UN's 1951 Geneva Convention and the 1967 Protocol, but it maintains the geographical limitation clause of the Convention and, therefore, only accepts applicants coming from Europe (Brewer and Yükseker 2009, 650). In the case of non-Europeans, this implies that the procedure to gain refugee status must be channeled through the UNHCR.[5] If granted refugee status, asylum seekers are resettled to other countries, mainly Australia, Canada, the United States, or Scandinavian countries; if it is denied, they have to leave Turkey. The metropolis of Istanbul, with its diverse urban districts and population, is a main hub for Congolese and other African migrants in Turkey. However, the living conditions for Congolese are difficult and precarious. Most Congolese do odd jobs, such as carrying boxes or goods for store owners or at the market, and others work in small workshops. Women sometimes find jobs as domestic workers or as sex workers.

Église Charité

Pastor Alain, who was introduced at the beginning of the chapter, had attended various revival churches and preached the gospel back in Kinshasa, the capital of the Democratic Republic of the Congo, for many years. However, it was only since his arrival in Istanbul that he had become a pastor, founding his own church in Istanbul in 2006. The Église Charité is one of around ten African Christian churches that have been founded since the beginning of the 2000s in Istanbul. Between twenty and one hundred churchgoers attend the weekly service of the church. Among them are asylum seekers, refugees, undocumented migrants, students, and suitcase traders,[6] predominantly from the DRC but also from other African countries in which French is spoken. The number of church visitors fluctuates depending on the mobility dynamics of migrants from the DRC to Turkey, as well as from Turkey to Greece or other Global North countries. Between three and ten newcomers appear at most Sunday services. At the same time, several older churchgoers have left the parish and Turkey altogether, having journeyed onward to Europe or other countries. At the end of each Sunday service, Pastor Alain asks newcomers to stay behind for a few minutes as he gives them an introduction to the main aspects of the church and life in Istanbul.

Faith and Space En Route

When Elisabeth, a church clerk, arrived in Istanbul in 2009, she didn't know anyone, as she explained to me in a conversation. In her search for fellow Congolese and spiritual assistance, she found and joined Église Charité, although she considered herself a Catholic. "This church here in Istanbul is very important for us Congolese. Well, you see, life here in Istanbul is difficult, and without the word of God it is even harder. In the church of Pastor Alain, I found things; the way he talked about the things that I did not know here, that helped me a lot, a lot! First and foremost, morally and spiritually, but then it is also like a community: We sing in Lingala, we sing in Swahili, and in French, we preach in French, and Pastor Alain preaches in Lingala." In our conversations, she narrated about everyday experiences of racism on the streets and financial precarity. Living as African Christians in a predominantly Muslim country, the church symbolizes, along with some other localities, a meeting point for the community and a space to regroup. Moreover, churchgoers underlined again and again the special style of spirituality, the "African" way of celebrating, preaching the gospel in Lingala and French and, most prominently, the particular music and dancing that allows them to experience the "power of the divine" during service.

Worshiping and Trading

Attending church on Sunday is commonly perceived as a Christian duty. However, as mentioned earlier, some of the parishioners also take advantage of another aspect of the church, namely by providing intermediary services for visiting traders, including accommodation assistance and facilitating trade within the informal business network of the church. After experiencing multiple wage frauds, Elisabeth also took this opportunity by joining the self-organized business network. On several Sundays, I was introduced to different Congolese traders coming from Vienna, Brussels, Kinshasa, Johannesburg, Abidjan, or Libreville who, during their stay, attended Église Charité. Most of these traders come to Istanbul to buy clothing or electrical devices or to order construction materials. Some of the traders I met are not only traveling to Istanbul but also making use of their connections, religious or otherwise, on business trips to Guangzhou or Dubai. Thanks to the constant coming and going of these traders, and the transiting and remaining church members, new transnational contacts were established and furthered within the church. This has, in turn, an effect on the local situation of the parishioners. At the end of 2012, Elisabeth was resettled to Canada and left Istanbul and the church. Before arriving at her new domicile, she had already contacted a Congolese revival church there, which she was planning to attend. In 2016, Pastor Alain was tired of waiting for resettlement. He made use of the 2013 newly introduced immigration law in Turkey, went back to DRC, and changed his status, obtaining a migrant residency permit, stating that he was no longer aiming to go to Europe. Being the head of the church and of his informal trading network, he explained to me that Istanbul would be more suitable for his religious as well as economic endeavors than any place in Europe.

Although Istanbul is primarily seen as a kind of transit hub for many migrants and refugees in the migration literature, in this research, I observed other features. As Église Charité is closely entangled with economic activities, its permanent fluidity was not only determined by ongoing transit but also by trading Congolese, who were the majority. Furthermore, some of my interlocutors, like Pastor Alain, decided to remain in Turkey and make use of certain aspects of the city. However, as the daily lives of a great part of the Congolese believers are still marked by social and religious exclusion, as well as precariousness, state arbitrariness, and racism, the church itself symbolizes a major place of refuge for the Congolese.

Rio de Janeiro—Living in and Next to the Favela

A relatively generous refugee law, launched in 1997, as well as comparatively open visa policies at the end of the 1990s, turned Brazil into an appealing transit and immigration country for Congolese (De Andrade and Marcolini 2002). Since 2003, the number of refugee applicants from the DRC has grown significantly (Tannuri 2010, 181). As in Istanbul, the majority of arriving Congolese plan to travel onward toward North America or Europe. Many of them have already transited South Africa or Angola to reach Brazil. Although there are no official numbers available, Congolese estimate the number of their coethnics in Rio to be around four hundred. A great proportion of them are refugees, undocumented migrants, or students. Arriving Congolese are predominantly young single women and men but also comprise families. Although most of them have middle-class backgrounds and many are graduates, the majority struggle to make a living in Rio as working hours are long, wages are low, and finding employment is difficult. Men typically work in construction, while women work as hairdressers, dishwashers, sex workers, or sell African dishes to the community.

"We were fleeing a war at home and we arrived here in Rio, and we find ourselves in a war again in our neighborhoods, in the favelas where we are living—a different war, the war between the drug dealers and the police," says Yves, a church member of Assembléia de Deus em Brás de Pina (Assembly of God in Bras de Pina), a Congolese revival church located in the northern zone of Rio de Janeiro, in one of our conversations during my first stay in 2011. Two years earlier, he had arrived in Brazil. His father had fled DRC in 2004, and it was in 2009 that he managed to bring three of his children, among them Yves, within the scope of family reunification. Though belonging to the upper middle class back in DRC, in Rio, they were residing in one of the disadvantaged neighborhoods in the city's northern zone.

In contrast to Istanbul, Rio's urban landscape is distinguishable for its thousands of Pentecostal churches. The percentage of Evangelicals and Pentecostals in the assumed largest Catholic country in the world rose from 6.6 percent to 22.2 percent between 1980 and 2010, with particularly high growth in Rio de Janeiro and São Paulo (IBGE 2012). This increase has been furthered by structures of urban violence, which for decades have dominated everyday life in the irregular settlements, the favelas. In light of this daily urban violence, certain religious interpretations, in particular those offered by the Pentecostal churches, became more conceivable and attractive to a rising number of favela inhabitants (Birman and Leite 2000, 278). Within this religious urban landscape, which predominantly spread throughout the favelas, there also exist various immigrant parishes. Contrary to their Brazilian Pentecostal counterparts, which openly agitate and direct their discourse toward the drug gangs dominating the favela territories, Congolese churches like the Assembléia de Deus em Brás de Pina do not intervene socially or politically within their neighborhood in order to not get involved in the "war" as Yves labeled it above. Though many of its parishioners live in favelas, the church decided to rent a space in a neighboring district. For many of its congregants, the church embodies a safe place in two ways: first, as a place that is spatially outside but still close to the favela and, thus, accessible, and second, as a space that is able to give their children both "cognitive strength" and provide a moral paradigm that might help to keep them away from drug dealing, alcohol, or drug abuse. As most Congolese migrants are not aiming to stay in Rio but to continue their journey, they tend to identify with a global, predominantly Congolese, religious network rather than their local environment.

Music Was My First Love

LK Kina, the head of the church music band *RDCongo Elembo*, had only spent a few weeks in Rio before our first encounter. Before reaching Brazil, he had lived in Angola and South Africa for about ten years, and he was planning to journey in the near future toward North America. Coming from a Baptist background, he had been singing in church since his early childhood. "Music was my first love, my first passion, and it is my vision! My desire is to become a professional singer," he told me in one of our conversations. While interviewing the members of the church music band, the entanglement between religious and global aspirations became apparent. Four of the seven members of the Congolese church band *RDCongo Elembo* were hoping to become professional musicians, and the church offered them the opportunity to rehearse and perform in public. Performing regularly in churches and cultural spaces and planning to record their first album, the musicians made use of Rio's significant evangelical music industry, which has existed since the 1990s (Oosterbaan 2006). Within the wide variety of Christian hip-hop, rock, and pop bands, they perform in French, Lingala, Swahili, and Portuguese, using Congolese and Angolan dance styles as well as Brazilian samba rhythms, thereby seeking to achieve recognition not only in Rio's music scene but also as African Christian musicians internationally. Their professional ambitions are not primarily associated with Rio specifically: the fact that the city offered them good opportunities to pursue musical careers did not constitute a reason to remain there. Most band members were preparing to continue their journey, whether to Europe, North America, or within Brazil.

In their research on African Pentecostal musicians in Israel, Sabar and Kanari (2006) have shown that some have managed to become well-known singers among African diasporic communities around the world. African religious music has gained enormous popularity as a commercial genre on a global scale (Devine 2011, 11). In the transnational market of music, literature, and videos circulating in global revival networks, performers do not have to rely solely on their success in specific locations. A few years after our initial encounter, LK Kina journeyed on to Canada with his family. Though not becoming a famous Christian musician, he today sings in his own church, which he founded in Toronto, gathering Congolese and other African revival Christians.

Upon closer examination, one can observe the intertwining of religion, diasporic networks, transnational mobility, and the pursuit of success. In the church setting, characterized by its transiting parishioners, the band makes music in order to praise God. At the same time, however, the musicians, following their aims of becoming professionals, make use of the vivid religious (and secular) music scene in Rio de Janeiro and mix African music styles with Brazilian rhythms. Meanwhile, they prepare for their next migration stage and hope to make use of the musical experiences they gained during their time in Rio for their next step, hoping that they will find another revival church with a band they could easily join. The existence of a diasporic religious music market gives them the confidence that they can further pursue their professional career in their prospective destination. This also points to the fact that finding refuge does not always mean that people do not aim to move further on. The reasons described above, such as urban violence and precarious working conditions, make people take journeys in the hopes of finding better living conditions.

Guangzhou—Going East

As in Istanbul, the Chinese trading city Guangzhou has seen an increasing presence of Congolese since the start of 2000, thanks to strengthened political relations between the two countries, easing travel. Furthermore, direct flights, the facilitation of visas, baggage allowances by travel agencies, and improved logistics have meant greater mobility (Braun 2019). While in 2000, the trade volume between China and the African continent amounted to 9 billion US dollars, in 2012, it was 166 billion USD (UN Comtrade 2014). Approximately 500,000 Africans were residing in China in 2012, with Guangzhou hosting some 100,000 (Bodomo 2020). According to Congolese community leaders, 450 Congolese were living permanently in Guangzhou in 2015, whereas a far greater number of Congolese (suitcase) traders frequently traveled back and forth not only between Guangzhou and Kinshasa but also between Guangzhou and Paris, Brussels, South Africa, Luanda, or Melbourne. These traders usually stay between three days and three months in the city. A large part of them are women, as they have increasingly become protagonists in the Congolese suitcase trade (Braun 2019).

Since China is considered a typical "non-immigration country," it manages its visa allocation very strictly. Most of the Congolese residents have established cargo agencies and hold business visas, which they have to renew every year, or hold a tourist visa. Others are students, many of whom also work at the same time as intermediaries for Congolese or other African businessmen. A few pursue careers as professional soccer players. Even Congolese who are married to Chinese spouses cannot get a stable residence status and have to renew their residence permit yearly (Niu 2018).

However, boosted by economic possibilities in the Guangzhou districts of Xiaobei and Sanyuanli, where many of the African traders and residents reside, African cargo agencies, restaurants, bars, hairdressers, and churches have emerged since 2000. Not all these churches are officially registered, but they are known by and under the surveillance of the Chinese authorities (Yang 2011). In 2015, eight Congolese revival churches existed in Xiaobei, some with smaller numbers of parishioners of up to ten churchgoers on Sundays, while others, like the Église de Puissance (church of power), are visited by around one hundred parishioners on Sundays.

"Living in China is like being on the battlefront," Felly Mwamba, the head of the Congolese community in China, to whom I was introduced during one of the church services at Église Puissance, tells me. At the time, Felly had lived for more than ten years as a businessman running a cargo company in Guangzhou. Before that, he had resided for several years in Dubai and Hong Kong, and he had traveled to Vietnam and Thailand to try his luck in business. Although leading a company with branches or contact offices in Brussels, Dubai, Kinshasa, Luanda (Angola), Nairobi (Kenya), and Johannesburg (South Africa), in the interview, he emphasized the precarious residence conditions even for businessmen, such as himself in China, and hence, the transient prospects of most Congolese in the city.

Église de Puissance

Located on the tenth floor in one of the largest commercial buildings in Xiaobei, the Tianxiu Building, the Église de Puissance is one of the eight Congolese churches that are said to exist in this Guangzhou neighborhood.

Whereas on the first floor of the building, retailers are selling clothing, electric devices, and "African drapery," on the tenth floor, the Sunday worship starts at 11 a.m. Around one hundred churchgoers are present at the worship every week. The church band, six women and five men, presumably all of them in their twenties, are leading with Christian pop music throughout the sermon. A Congolese pastor and trader from Kinshasa who regularly visits Guangzhou is preaching: "If you are not in dialogue with God, and you are coming to Guangzhou to do business, you will not be successful. You will buy the same T-shirts as all the others do, and you will return to Kinshasa and not be able to sell them. Though you might be praying, you may not reach the right dialogue with God. Therefore, you do not really listen to his advice, and you are buying the wrong articles."

Preaching on prosperity and economic and social success is a widespread characteristic of Congolese revival churches around the globe. However, in the Église Puissance, the content of the sermons is closely adjusted to the parishioners' reality and the daily business in the local markets. Accordingly, pastors not only preach the word of God, but their services also instruct the churchgoers on how to yield profit. In this context, the aspiration of moving on, in an economic or geographical sense, fits with new social patterns, practices, and networks of this religious movement. "Our daily life is surrounded by business," explains Camille, one of the music band members of Église Puissance, "so, apart from worshiping, we are interested in business, this is our life here in China." Also, for Pastor Josef, the head of the church, being a "man of God" is accompanied by being a businessman. During his frequent business trips to Brussels, Paris, and Kinshasa, he preaches the gospel in various churches. In turn, pastors pay him visits back to preach at the Église de Puissance in Guangzhou while they are in the city for their trading activities.

Similar to such churches in Istanbul, the church operates as a connecting hub facilitating the mobility of its pastors and congregants. However, in contrast to the situation in Istanbul, the primary aim of most parishioners is trade. Among them, many do petty trade between Guangzhou and Kinshasa or Luanda, like Helena Okito. Through a pastor of her church back in Kinshasa, she embarked for Guangzhou for the first time in 2012 and has since prolonged her visits from one week to several months. Predominantly purchasing clothing at the cheap wholesale outlets on the outskirts of the city, she has, over the years, enlarged her business activities and become a broker for other Congolese traders. Elaborating on her business strategy, she emphasizes the importance of time and place. "You need to know where to be and at which time. For instance, the earlier you reach certain markets, the cheaper you can purchase the commodities." In this vein, attending church on Sunday mornings, which she interprets predominantly as a Christian duty, is a way to gather with other Congolese but also to potentially meet new clients.

Conclusion

In a conversation about increasing global mobility restrictions in Kinshasa in 2010, a Congolese pastor concluded the conversation by stating, "The Gospel doesn't know borders, neither do we, so we have, and we will always move. Borders do not stop the people." Following this quote, I have laid out in this chapter the migration and mobility strategies of Congolese revival Christians in three Global South metropolises. Being mobile, they have globally expanded a "religious infrastructure" in which believers share religious and day-to-day life experiences and keep up with their religious duties, which serve as reference points. In this infrastructure, which is based on and emerged through the mobility of the believers themselves, they take either uni- or multidirectional routes and make detours, remaining for longer or shorter periods in different cities. While expanding, they have not only taken root but have also contributed to the transformation of the physical and social landscapes in the cities, in this case, Istanbul, Rio de Janeiro, and Guangzhou. Consequently, new parishes with new profiles have been launched in different locations of migration and have aligned to the settings of the respective society. Following AbdouMaliq Simone's reasoning, one could also see the sophisticated and widely ramified network as a process of "worlding from below," reaching out to a "larger world" through "circuits of migration, resource evacuation and commodity exchange" (Simone 2001, 15). Through transnational agency, Congolese revival Christians participate locally in the permanent production of the city, a city which is not defined as a circumstanced territorial unity but rather as a globally connected assemblage of entangled religious, political, and social practices, spaces, symbols, and materialities.

Against the background of this flexible adjustment in the three metropolises that I presented, urban space, as well as national structures, had an impact on both the "sacred spaces" and the functions those spaces fulfilled for the believers. Ananya Roy and Aihwa Ong (2011), as I introduced earlier, suggest "worlding" as an approach to understanding how urban configurations are both locally specific and embedded in global processes and networks. This allows us to trace the dimensions and forms in which a network of Congolese revival Christians expands globally while adapting differently to urban and national settings.

The entanglement of religious, migratory, and economic aspirations within the everyday life of the believers, as well as in the church itself, becomes apparent. As we can discern in the statements of Pastor Alain, religious ambitions are used as justification for migration, as well as for eventual anchoring. Here, as in the context of Guangzhou, religious, migratory, and economic aspirations go hand in hand. It once again becomes apparent that the clear distinction between the economic aspects of migration and the cruelties of forced migration are interlinked. Use of these distinctions, which are predominantly conceived by policymakers, is not sustainable, as the large part of my interlocutors were nothing and all of it at the same time: asylum seekers, refugees, residents, and traders.

One facet that makes Pentecostalism so attractive in the migration process is that it offers believers the opportunity to deal with the experience of social change, i.e., continuity and discontinuity, and supports adaptation to a changing environment. In addition, the special nature of the service, the unique style, and the music with its sensational affinity (Meyer 2009) give parishioners a special spirituality from which they draw strength to face the hardships and uncertainties they encounter in their daily lives at various levels.

Religion is one of the many vehicles that act as an interface for Congolese migrants in cities around the world. In the metropolises studied here, the settlement of religious, social, and entrepreneurial projects can be observed as they are broadly related to a globally expanding religious commitment. In this sense, religion is a reference point that connects the initiatives, styles, interpretations, and experiences of believers themselves, which form the basis on which believers create a "religious infrastructure." This infrastructure supports their mobility and also offers them, even in conditions of discrimination and exclusion, a wider range of opportunities.

Notes

1 This article is dedicated to Pastor Alain Suku, who generously supported me in my research in Istanbul and beyond. We frequently had lunch together; he invited me to his home and to accompany him in his daily life and business. We discussed Congolese politics, the situation of Congolese migrants in Istanbul, the role of the church, his own ambitions, and beyond. Pastor Alain Suku passed away on November 21, 2020, in Istanbul.

2 Lingala is one of the main languages spoken in the Democratic Republic of Congo.

3 This chapter is based on findings from the research project "Migrants' Routes–Religious Routes: The Role of (Neo-)Pentecostal Churches in the Migration Routes of Congolese Migrants," which was conducted in the scope of the research project "Global Prayers: Redemption and Liberation in the City," as well the research project "Entrepreneurial Citizenship: New Economies and the Respatialisation of

Citizenship in China, India, and Europe." The interviews and conversations I use in this chapter were conducted in French and subsequently translated into English by me. The names of most of the interviewees have been changed.

4 As in other African contexts, it is difficult to maintain the differentiation between the so-called Églises de Réveil and Pentecostal churches in Kinshasa (Pype 2006, 299). Since almost all my interviewees defined their parish as an Église de Réveil, I am using the term *revival church* in this chapter.

5 This geographical limitation to European citizens for full recognition of refugee status is also retained in the Law on Foreigners and International Protection (LFIP) that came into effect on April 12, 2014.

6 Suitcase trade is a term generally used to describe transnational unregulated and unregistered commerce.

References

Alpes, Maybritt Jill. 2017. "Why Aspiring Migrants Trust Migration Brokers: The Moral Economy of Departure in Anglophone Cameroon." *Africa: Journal of the International African Institute* 87 (2): 304–21. http://www.jstor.org/stable/26158171.

Anderson, Allan. 2014. *An Introduction to Pentecostalism: Global Charismatic Christianity.* 2nd ed. Cambridge: Cambridge University Press.

Awad, Ibrahim, and Usha Natarajan. 2018. "Migration Myths and the Global South." *The Cairo Review of Global Affairs* 30: 46–55. https://www.thecairoreview.com/essays/migration-myths-and-the-global-south/.

Bauman, Zygmunt. 1998. *Globalization: The Human Consequences.* New York: Columbia University Press.

Birman, P., and M.P. Leite. 2000. "Whatever Happened to What Used to Be the Largest Catholic Country in the World?" *Daedalus* 129 (2): 271–90. http://www.jstor.org/stable/20027637.

Bodomo, Adams. 2020. "Historical and Contemporary Perspectives on Inequalities and Well-being of Africans in China." *Asian Ethnicity* 21 (4): 526–41.

Braun, L. 2019. "Wandering Women: The Work of Congolese Transnational Traders." *Africa* 89 (2): 378–97. doi: https://doi.org/10.1017/S0001972019000135.

Brewer, Kelly Todd, and Deniz Yükseker. 2009. "A Survey on African Migrants and Asylum Seekers in Istanbul." In *Land of Diverse Migrations: Challenges of Emigration and Immigration in Turkey,* edited by Ahmed Içduygu and Kemal Kirişci, 637–720. Istanbul: Bilgi Üniversitesi Yayınları.

Coleman, Simon. 2000. *The Globalisation of Charismatic Christianity: Spreading the Gospel of Prosperity.* Vol. 12. New York: Cambridge University Press.

Cranston, Sophie, Joris Schapendonk, and Ernst Spaan. 2018. "New Directions in Exploring the Migration Industries: Introduction to Special Issue." *Journal of Ethnic and Migration Studies* 44 (4): 543–57.

De Andrade, José H. Fischel, and Adriana Marcolini. 2002. "A política brasileira de proteção e de reassentamento de refugiados: breves comentários sobre suas principais características." *Revista brasileira de política internacional* 45 (1): 168–76.

De Boeck, Filip, and Jean-Pierre Jacquemin. 2006. "La ville de Kinshasa, une architecture du verbe." *Esprit,* December 12, 79–105.

Devine, Kyle. 2011. "The Popularity of Religious Music and the Religiosity of Popular Music." *Scottish Music Review* 2 (1): 1–22.

Garbin, David. 2010. "Symbolic Geographies of the Sacred: Diasporic Territorialization and Charismatic Power in a Transnational Congolese Prophetic Church." In *Traveling Spirits: Migrants, Markets and Moralities*, edited by Gertrud Hüwelmeier and Kristine Krause, 145–64. London: Routledge.

Garbin, David, and Manuel Vasquez. 2011. "God Is Technology: Mediating the Sacred in the Congolese Diaspora." In *Migrations, Diaspora and Information Technology in Global Societies*, edited by Leopoldina Fortunati, Raul Pertierra, and Jane Vicent, 157–71. London and New York: Routledge.

Heck, Gerda. 2014. "Worshipping in the Hotel Golden Age: Transnational Networks, Economy, Religion and Migration of Congolese in Istanbul." In *Global Prayers: Contemporary Manifestations of the Religious in the City*, edited by Jochen Becker, Kathrin Klingan, Stephan Lanz, and Kathrin Wildner, 274–89. Zürich: Lars Müller.

Honwana, Alcinda Manuel, and Filip de Boeck, eds. 2005. *Makers & Breakers: Children & Youth in Postcolonial Africa*. Oxford, Trenton, Dakar: James Currey.

Hüwelmeier, Gertrud, and Kristine Krause, eds. 2010, 2009. *Traveling Spirits: Migrants, Markets and Mobilities*. Vol. 4. New York: Routledge.

IBGE–Instituto Brasileiro de Geografia e Estatística. 2012. "Censo 2010: número de católicos cai e aumenta o de evangélicos, espíritas e sem religião." https://censo2010.ibge.gov.br/noticias-censo.html.

Kalu, Ogbu. 2010. "African Pentecostalism in Diaspora." *PentecoStudies* 9(1): 9–34. https://doi.org/10.1558/ptcs.v9i1.9.

Levitt, Peggy. 2007. *God Needs No Passport: Immigrants and the Changing American Religious Landscape*. New York: The New Press.

Lin, Weiqiang, Johan Lindquist, Biao Xiang, and Brenda S.A. Yeoh. 2017. "Migration Infrastructures and the Production of Migrant Mobilities." *Mobilities* 12 (2): 167–74.

Marcus, George E. 1995. "Ethnography in/of the World System: The Emergence of Multi-Sited Ethnography." *Annual Review of Anthropology* 24 (1): 95–117.

Meyer, Birgit. 2004. "Christianity in Africa: From African Independent to Pentecostal-Charismatic Churches." *Annual Review of Anthropology* 33 (1): 447–74.

———. 2009. "From Imagined Communities to Aesthetic Formations: Religious Mediations, Sensational Forms, and Styles of Binding." In *Aesthetic Formations: Media, Religion, and the Senses*, edited by Birgit Meyer, 1–30. New York: Palgrave.

Niu, Dong. 2018. "Transient: A Descriptive Concept for Understanding Africans in Guangzhou." *African Studies Quarterly* 17 (4): 85–100.

Ong, Aihwa. 2011. "Worlding Cities, or the Art of Being Global." In *Worlding Cities: Asian Experiments and the Art of Being Global*, edited by Ananya Roy and Aihwa Ong, 1–26. Malden, MA and Oxford: Wiley-Blackwell.

Oosterbaan, Martijn. 2006. "Divine Mediations: Pentecostalism, Politics and Mass Media in a Favela in Rio de Janeiro." PhD diss., University of Amsterdam. https://dare.uva.nl/search?identifier=ffa3b10d-a09e-4b24-a22d-4b432d3613ab.

Pype, Katrien. 2006. "Dancing for God or the Devil: Pentecostal Discourse on Popular Dance in Kinshasa." *Journal of Religion in Africa* 36 (3–4): 296–318.

Roy, Ananya. 2009. "The 21st-Century Metropolis: New Geographies of Theory." *Regional Studies* 43 (6): 819–30.

Roy, Ananya, and Aihwa Ong, eds. 2011. *Worlding Cities: Asian Experiments and the Art of Being Global.* Malden, MA and Oxford: Wiley-Blackwell.

Sabar, Galia, and Shlomit Kanari. 2006. "'I'm Singing My Way Up': The Significance of Music Amongst African Christian Migrants in Israel." *Studies in World Christianity* 12 (2): 101–25.

Schoumaker, Bruno, and Marie-Laurence Flahaux. 2013. "Changing Patterns of Congolese Migration." *MAFE Working Paper No. 19.* Paris: INED. https://www.ined.fr/fichier/s_rubrique/22089/wp19_.fr.pdf.

Simone, AbdouMaliq. 2001. "On the Worlding of African Cities." *African Studies Review* 44 (2): 15–41.

———. 2004. "People as Infrastructure: Intersecting Fragments in Johannesburg." *Public Culture* 16 (3): 407–429.

Tannuri, Maria Regina Petrus. 2010. "Refugiados congoleses no Rio de Janeiro e dinâmicas de 'integração local.'" PhD diss., UFRJ.

UN Comtrade. 2014. United Nations Commodity Trade Statistics Database, SITC Rev. 3. https://comtrade.un.org.

van Dijk, Rick. 2012. "Pentecostalism and Post-Development: Exploring Religion as a Developmental Ideology in Ghanaian Migrant Communities." In *Pentecostalism and Development: Churches, NGOs and Social Change in Africa*, edited by Dena Freeman, 77–108. London: Palgrave Macmillan.

Vásquez, Manuel A., and Kim Knott. 2014. "Three Dimensions of Religious Place Making in Diaspora." *Global Networks (Oxford)* 14 (3): 326–47.

Xiang, Biao, and Johan Lindquist. 2014. "Migration Infrastructure." *The International Migration Review* 48 (1): 122–48.

Yang, Yang. 2011. "African Traders in Guangzhou: Why They Come, What They Do, and How They Live." PhD diss., The Chinese University of Hong Kong. https://core.ac.uk/download/pdf/48537913.pdf.

11

Sounds of Transit Migration: Chaldean-Iraqi Migrant Music in Istanbul

Evrim Hikmet Öğüt

Introduction

The transit migration experience differs from permanent migration according to various factors such as legal status uncertainties, the size of the community, availability of economic opportunities, and psychological awareness of the temporal nature of the community's transitional existence within the geographic/national transit stop.[1] These differentiations are apparent in the production and the use of cultural forms, including music. Music reflects and creates social relations; it constitutes a ground for understanding the migratory experience and, thus, the specific circumstances and awareness of impermanence in transit migration. This chapter, based on long-term field research between 2011 and 2014, focuses on the specific characteristics of the transit migration experience, illustrating some basic aspects of the Chaldean-Iraqi migrant community's musical practices in Istanbul.

Transit migration is a common phenomenon for various countries on migratory routes from Middle Eastern, Asian, and African countries to Europe and other Western countries. During recent decades, thousands of migrants from these regions transited and entered neighboring countries seeking asylum. Turkey, Greece, and Italy, among other countries, played roles as the gates and gatekeepers of "Fortress Europe."

As a country that does not give refugee status other than to applicants from Europe,[2] Turkey has operated as one of the application centers for asylum seekers and economic migrants for the past few decades. Even though Turkey's position has changed to that of a host country since 2016, primarily for Syrian migrants, and in effect operates as an external border of the European Union, it still resists permanent migration policy solutions. Besides Syrians, migrants from Iran, Iraq, Afghanistan, and various other Asian, African, and Middle Eastern countries enter Turkey regularly or irregularly to pass or be sent to a third country. Among these is the Chaldean-Iraqi transit community.

Iraqi migratory patterns include various religious and ethnic groups that accelerated after the 2003 US invasion. Iraqis constitute a substantial proportion of all migration from the Middle East to Turkey. Chaldeans, the most populous Christian community in Iraq, are among the major actors in this migratory movement, using Turkey as a transit country prior to their arrival in a third country through resettlement.

Chaldean-Iraqi Migration to Turkey

Chaldean migration to Europe and other Western countries began with departures from Iraq, the ancient homeland of the community, in the late nineteenth century mostly due to economic reasons (Albayrak 1997, 150; Sengstock 2005, 3), and continued with Chaldeans living in the Ottoman territory, along with other non-Muslim populations forced to migrate in 1915.

Iraqi emigration accelerated during the 1980s due to the Iran–Iraq War and then the Gulf War in 1991, reaching its peak with the US invasion of Iraq in 2003. In 2006, the United Nations High Commissioner for Refugees (UNHCR) took the advisory decision that asylum requests from the center and south of Iraq should not be denied due to ongoing violence in these regions. After this decision, the number of Iraqi migrants in Turkey reached 10,000 (Danış 2010, 6); Chaldeans, the largest Christian minority in Iraq,[3] constituted a large portion of this flow. Finally, ISIS's entry into Mosul in 2014 brought another wave of Iraqi asylum seekers to Turkey.

Members of the Chaldean community mentioned in this study consist of migrants who left Iraq after 2003, are in different stages of the UN application process, and live in Turkey, mostly under asylum seeker status. Due to Turkey's immigration and asylum policy, the migrants residing in Turkey as transit migrants before settling in a third country, such as the USA, Canada, or Australia are expected to live in one of sixty-two migrant satellite cities (such as Sinop, Van, Bilecik, Yalova, and so on); however, the lack of Christian religious networks and the scarcity of job opportunities in these cities lead them to look for ways to stay in Istanbul.

Chaldean-Iraqi Migrants in Istanbul

Istanbul hosts Christian religious organizations and networks that provide broader cultural and economic opportunities for Chaldean-Iraqi migrants than can be found in satellite cities. Thus, it has become a center where Chaldean-Iraqis have congregated, as other temporary migrant communities have also done.

As temporary residence in Turkey can vary from six to eight months to a few years, temporary migrant communities depend on migrant networks that enable them to maintain their daily lives, find accommodation and work, handle application procedures, and satisfy basic needs like education and healthcare. The literature on migration points to the importance of social and institutional migrant networks, formed by those who migrated earlier, for the decision to migrate, to minimize cost and risk, and to resume daily life in the migrated country (Massey et al. 1993, 449; Sassoon 2010, 35; Haug 2008, 590; Faist 2004). In the specific case of Iraqi Christian migration to Turkey, kinship and religious networks emerge as especially significant (Danış 2007, 605–606). There is also a Chaldean-Assyrian community in Turkey, which constitutes a part of the networks mentioned above, and the associations serving this community and international Catholic charity organizations constitute the institutional branch.

The importance of religious networks for the Chaldean-Iraqi community in Istanbul explains why the primary locations of this field study were religious institutions such as St. Esprit Church and the Don Bosco Foundation Hall, where young community members gather during weekends.[4]

The Chaldean-Iraqi youth who attend the weekly gatherings at the Don Bosco Foundation Hall and sing in the church choir are the focus of this study. These young people do not have the opportunity to continue their education in Turkey due to the language barrier. During their stay in Istanbul, they remain unemployed or hold low-paid, flexible, and precarious jobs without a work permit. Domestic work, work at small textile mills, and work as a porter at these mills are among the jobs held by Chaldean youth temporarily residing in Istanbul.

Temporary Migration Experience, "Waiting"

Even though transit migration has a history back to the Second World War, the term today is mostly used to refer to actual or potential irregular migration in the European context (Düvell 2006). When we consider the variety within human mobility, the possible continuation of any specific migratory movement, and the reality that the migration experience can transform due to changing legal status, the terms "temporary" and "permanent" migration emerge as somewhat ambiguous descriptions. For instance, in the case of Syrian refugees in Turkey, most hope to leave as soon as possible for a European country. However, due to the limitations imposed on their crossing to Europe since March 2016,[5] even though they hold a "temporary protection" status, their stay has often become permanent.[6]

On the other hand, it is evident that temporary or "guest" status in a given location profoundly affects both the emotional experience of migrants and daily life practices; thus, guest status plays a decisive role in the migratory experience. The Chaldean-Iraqi community has internalized its "temporary status," mostly due to third-country high acceptance rates as required by a UNHCR advisory decision. Hence, they are likely to believe that they will be resettled in a third country soon. Their relatively limited relationship with the local culture and unwillingness to learn Turkish presents compelling evidence for a specific state of mind associated with being temporary migrants.

In addition to the main problems emerging from a tenuous legal status, precarity, difficult working conditions, and other obstacles due to religious and linguistic differences, the temporary migratory experience also displays a strong sense of uncertainty. "Limbo" is a highly used term in transit migration studies, and the notion of "waiting," mostly discussed in the migratory context related to asylum-seeking and refugee experiences (Wong 1991; Conlon 2007; Pascucci 2015), is very crucial in the transit experience.

The specific character of transit migration extends the dual temporality of the migratory experience beyond the past and present by adding another: the future. Jonathan Shannon states that for Syrian migrants in Istanbul, the spatial-temporal dimension of migratory experiences and "how they saw their past, present, and future horizons . . . offers important insights into how they imagine and recreate home in conditions of displacement" (2019, 2175). Similarly, in the Chaldean-Iraqi context, the home does not exist in the past, nor can it be built here. The psychological impact of being in transit on young migrants reveals itself in two, sometimes intertwined, ways: Since the young are hopeful about their future, they are ready to tolerate difficulties in their first few months. Nevertheless, with any problems in the legal process, extended waiting periods, or the need to apply to another country due to a rejection from the previous one, they start to experience an overwhelming sense of being trapped in the transit experience.

Ayhan Kaya (2002, 59) argues that the diaspora identity is built on two contradictory dimensions. Considering cultural contexts, it can be said that there are two cultural behaviors related to these, which may be seen simultaneously. First, the particularist axis accentuates preserving national or ethnic identity, and second, the universalist axis negotiates with "the other" in its interactions. The former mostly emphasizes cultural practices related to past experiences, and its spatial reference is to the homeland. These practices are marked by concepts like nostalgia, collective memory, protection of the culture, etc. The latter axis is related to the present, and its spatial reference is to the destination country. These practices seek change and an exchange in which new and hybrid practices may emerge (Bohlman 2001).

However, since the transit country is not the final destination for migrants in transit, the future—near or far, threatening or hopeful—is always a part of their lives and thoughts. Thus, I suggest a third spatial dimension related to the future and the final destination of the transit migratory experience. Even though I suggest employing this triple temporal categorization as an analytical tool for the community's musical practices, the third component is only visible in the plans and hopes of the young migrants, such as studying music and forming bands with friends who have already migrated to a third country. Considering this chapter's limitations, it will suffice to give some examples in the following sections.

Transit Migration in Ethnomusicology

Migration and music studies developed rapidly at the beginning of the twenty-first century. However, in this broad literature, temporary migration experiences are mostly neglected (Öğüt 2015b) or otherwise usually discussed within the context of individual migration (Beckerman 2010; Bohlman 2010; Scheding 2010; Reyes 1999; Kaiser 2006; Landau 2011). One of the main reasons for this is the difficulty of conducting long-term field research with a community that is characterized by mobility and circulation. Secondly, it is believed that, because of the limits imposed on musical practices by transit circumstances, from having professional performers or musical institutions to being able to organize concerts, the transit experience may provide limited data for a researcher. However, another important reason for ignoring transit communities in music research is that these studies are mostly conducted in Western countries, which are usually the destinations rather than the places of transit.

Diaspora is one of the primary concepts shaping ethnomusicologic migration studies (Solomon 2014, 319), but the Chaldean-Iraqi community's status in Istanbul as a diasporic community is unclear. Chaldean-Iraqi migrants in Turkey have connections, though limited, with community members' transnational networks in the homeland and other countries. However, the community's temporary nature hinders the establishment of permanent economic relations and institutions, preventing the community from having more permanent educational institutions, associations, and establishments. In addition, under the circumstances, community members cannot take an active political role in the destination country or homeland. Regarding common features associated with diaspora communities,[7] it became difficult to understand the Chaldean-Iraqi community in Istanbul in the frame of diaspora theories. Therefore, while giving attention to the transnational aspect of the community's musical activities, I avoid using the diaspora concept.

When we examine the musical practices of the Chaldean-Iraqi community in Istanbul through different aspects, such as religious and nonreligious music, amateur and professional performance, and the preference for being a performer or listener, the most notable characteristic of this vibrant music scene is that these practices carry the traces of temporary migration experience. In this context, this essay examines the effects of temporary migration experience on musical practices—a subject barely studied in the field of ethnomusicology—in contrast to the effects of permanent migration on music production, performance, and consumption. Therefore, this chapter also proposes a novel contribution to migration studies in ethnomusicology.

Chaldean-Iraqi Musical Practices in Istanbul

Despite the small size of the population and the limitations of transit conditions, music arguably has an even more prominent role in various daily activities and special events of Chaldean-Iraqi migrants than would normally be the case. The community's musical practices include various genres and social contexts, from religious musical practices to listening habits. Considering the limits of this chapter, I will only deal with the examples of some of those in the context of transit experience.

Religious Music

The church choir performances and the hymn repertoire performed in weekly or special services constitute the central feature of the Chaldean-Iraqi community's religious musical practices in Istanbul. Since it is possible to examine religious music from various perspectives, I will provide a specific case to illustrate the role of knowledge of religious music for social acceptance.

For the Chaldean-Iraqis in Istanbul, the church is more than a religious institution, as it is the primary place of focus for community interaction. It is where the community members come together and share food and refreshments in the courtyard, accompanied by conversation after Sunday services. Because of the importance of the church, participation in the church choir plays a crucial role in incorporating the newly arrived young members into Chaldean-Iraqi community networks.

The story of my interlocutors, Flona, Farah, and Manuela, exemplifies music's role in their acceptance in the community. These three sisters arrived in Istanbul in 2011 with their mother and brother. When they noticed no permanent choir performance in the religious services, as well-educated choir singers in their homeland, they prepared a hymn repertoire especially suited for the Feast of the Cross in the following month and took over the role of the church choir. As their attendance enriched the service, which had lacked a choir performance for a long time, their performance was appreciated by the community members and church officials, leading them to be invited as permanent members of the choir. Ensuring choir performance in services, hence filling a critical gap, these young members gained the community's respect in a short time. Soon after, they became the informal leaders of the choir and even invited their relative, Jan, to play the keyboard during the services. Jan's comment, which does not regard performing in the church as a source of income, describes well the advantage of being a church musician and gaining social recognition in the eyes of community members:

> If I had not played in the church, I could be an assistant to the priest. I received religious education for seven to eight years; my brother is a priest in France. However, performing in the church enabled me to meet more people and gain a reputation. If I had become an assistant, they could recognize me, but now more people know me since I am playing the keyboard. (Personal interview, May 21, 2012)

I find both examples, of the three sisters and Jan, significant in terms of demonstrating that arriving in a new country with a musical background and skills and command of church repertoire enabled, following Bourdieu (1986), the conversion of cultural capital into social capital that has been partially lost in the migration process.

Soon after starting my field research and introducing myself to the choir members, I was invited to play the violin at the church as an accompanist. Playing at the church's weekly services, I gained community members' trust, thereby opening doors for me and positively transforming my field research process. Therefore, as an outsider, my experience was somewhat intertwined with my interlocutors, and the ensuing welcoming attitude toward my research constitutes a striking example illustrating the significance of being a church musician to be accepted by the community.

My experience also presents the opportunity to discuss community tactics to ensure the survival of rituals during temporary migration. During the years I intermittently accompanied the choir, I was worried that, as a musician trained in Western classical music and without expertise in Chaldean music[8] or *maqam* music[9] performance in general, I might cause "damage" to the church repertoire. Indeed, at the end of my field research, when I was transcribing the recordings of several hymns from rehearsals and rituals, I noticed a distinct difference between the recordings of the same hymns, the version accompanied by me and the other without me. I realized that in the performances where I played the introductory melodies with subtle rhythmic mistakes, the choir repeated those musical elements in the same manner without any correction.

Their tolerance of variations to such a delicate and "absolute"[10] repertoire as religious music highlights the fact that their need for my accompanying performance served as the determining factor for this flexibility. Mary C. Sengstock's research (2005) on Chaldean life in Michigan, a state with a major Chaldean population, indicates that the community is tightly organized around tens of churches, established institutions, economic networks, and religious musical practice maintained by experienced members. In this regard, it is obvious that a non-Christian musician like me, someone outside the community who is unfamiliar with the music tradition, will not be welcomed to the church, neither in the homeland nor in the countries with a permanent Chaldean diaspora. However, in Istanbul, the transit location of temporary migration experience, the community prioritizes the continuance of religious musical practices, and any temporary support is much appreciated.

Lastly, it is also possible to observe the traces of temporary migration in the overall Chaldean-Iraqi hymn repertoire in Istanbul. While one could expect to see a very diverse repertoire from a choir made up of members from different parts of Iraq, only a few members are in a position to lead the choir due to the beauty of their voice or their knowledge of music/repertoire. Thus, the repertoire in rituals remains limited to the pieces within this smaller group's expertise.[11] Even though it is not possible to cover the sources and the surprising liveliness of this repertoire here, I will also mention a newly composed hymn in the context of the transnational musical circulation in the following sections.

Nonreligious Music

Chaldean-Iraqi youth gather at the Don Bosco Foundation every Saturday afternoon and socialize with each other; they play games such as football, table tennis, basketball, and volleyball, listen to music together, dance together, and sing on special occasions such as Easter, Christmas, and other holidays. These gatherings are among the main social activities of the young community members.

Any gathering at Don Bosco that is accompanied by music from computers or mobile phones, without exception, is concluded with *dabkeh* dance[12] and collective praying. As mentioned by the young members, this kind of dance is only performed on special occasions and at weddings in Iraq. In this respect, it is noteworthy that these line dances become more routine during the migration process. Landau (2011, 39) states that it is common for music and dance practices to gain new meanings with the migratory experience. In the case of *dabkeh,* the dance no longer marks special occasions but rather becomes part of the daily routine. We can also consider this transformation in the context of cultural memory in which ritualistic events gain new forms and new rituals are invented (Assmann 2011). Moreover, taking into consideration that the dance is followed by collective praying, it appears that these collective dance practices have a unifying and healing impact on the community.

Young male members are eager to sing solo performances at the parties organized for special occasions in the Don Bosco Foundation Hall, since there are almost no musical instrumentalists in the community or those who can play were unable to bring their instruments to Istanbul with them.[13] Musical performances at the events consist of singing on stage, with a microphone, and without accompaniment. Young women, who do not usually join these performances, prefer to sing hymns as a chorister or soloist in the church choir.[14]

The dominant performance repertoire of the community gatherings consists of popular traditional folk and urban music and *mawwals.*[15] Even though I have heard some other popular genres—both in Arabic and Turkish—in the repertoires of these vocal performances, it is worth noting that those cases were quite scarce. The repertoire mostly comprises tunes that emphasize the quality of voice.[16] *Mawwal* is the preferred genre to prove one's voice quality and singing ability. Most of the songs are about love for the homeland, longing, displacement, and migration.

It should be noted that Chaldean-Iraqi youth listen to a wide range of music in Istanbul, as is the case in the homeland and third countries. These young people, who are Chaldean, Iraqi, migrant, Arab, and Christian, all at the same time, and young individuals in the globalized world, listen to Arabic, Chaldean, Turkish, and English popular music, just as many of their international peers listen to different types of music. Similarly, they share video clips from all these types of music on their social media accounts. However, when it comes to singing, they mostly prefer the songs from their region in the Chaldean-Assyrian, Arabic, or Kurdish languages. This preference indicates an overlap with the listening habits of their parents' generation. Likewise, my young interlocutor Yusuf's mother and father ask him to sing "sad" songs to them, which makes them cry, and Rami's mother, Nazdar, sings sad songs at home all the time after the loss of her spouse. Thus, while their singing practices refer to past experiences and feelings about the past, their listening habits refer to the present. Similarly, the former link them to their community, while the latter connect them to a broader group of young people around the world.

In terms of contemporary genres, rap music makes a considerable contribution to young male members' performances. Interestingly, the "traditional" or regional repertoire is not in conflict with these rap songs. They mostly coexist: a singer can start with a *mawwal*, a free rhythm improvisation based on a traditional musical pattern, as an introduction and continue with a rap song, or a *mawwal* performance of one singer can constitute a musical background for another singer's rap performance. These performances do not require a musical instrument and can be performed with friends beatboxing and tapping, paving the way for creativity. This is also why rap music and hip-hop culture are widely used in various migratory experiences (Kaya 2002, 43). While the lyrics mostly depict migration, violence in the homeland, and longing for family members, language preferences vary depending on thematic associations. Regarding a rap song Ricky—a young member who came from the Kurdish region in Iraq—himself composed in four languages about the various phases of his journey, his feelings about his experience, and his multiple identities, he elaborated:

> In the first part in Arabic, I talk about terrorists in my country who only know about killing people and their darkness. The second part is in English; it talks about how I came to another country and became a "foreigner." In the third part, I repeat the sense of being a foreigner in Assyrian and say that we should fight for our country. The last part is in Kurdish; I talk about my family in Erbil, how my mother said that I abandoned them, and how much I miss her. (Personal interview, Ricky, September 9, 2014)[17]

There was constant migration flow from 2006, both from Iraq to Turkey and from Turkey to third countries. Due to this increasing circulation, the small size of the community, and the difficulty of carrying an instrument in the migratory process, in the Chaldean-Iraqi community in Istanbul, regular performances from professional musicians are lacking. For special occasions, the community relies on performances of young DJs from the community rather than live music. Danyal, one of the two Chaldean DJs I met during my study and who later left for the US, performed as a DJ at community events such as birthday parties of young members, weddings, engagements, and other special events during his stay in Istanbul. When he left Istanbul to settle in Detroit, USA, he was unable to take his mixer and, therefore, left it with his cousin for community use. Similarly, Rafi left some of his playlists before his departure. As Manuela complained, during his absence, the community was not pleased with the performance of Turkish DJs:

> In fact, the songs he [the Turkish DJ] played were good, but we have other songs. Rafi played at the parties here; he was a professional DJ who performed at weddings in Iraq. He was playing for free at the parties but took money at weddings. Parties were much better when he was here. He left some music before he went, but unfortunately, it is not much. (Personal interview, Manuela, January 4, 2013)

Transnational Musical Practices

With the lack of professional musicians and musical instruments, digital technologies, such as listening to music via computers or smartphones and DJ performances, gain a significant place. As Benedict Anderson (1994, 322) argued, as early as 1994, the technological revolution profoundly affected the immigration experience. Smartphones and sharing platforms on the internet are having a massive impact on migrants' lives across social classes. Even the poorest households have satellite receivers to watch TV channels broadcasting from Arab countries (Danış 2006, 8). Studies focusing on Iraqi migration reveal that those who were able to leave the country were mostly middle and upper-middle-class individuals or families.[18] Even though the migration process leads to a noticeable decline in these individuals' economic status, there appears to be no change in their smartphone and internet consumption habits. These technologies have become indispensable tools for Chaldean youth in Istanbul to access transnational networks.

Chaldean migrants in Istanbul, similar to many other migrant communities, communicate with relatives in the homeland and other countries via communication tools. One may assume that this desire stems from the community's conservative attitude regarding preserving family relations and how these family relations are considered one of the main unifying factors in the community (Gallagher 1999, 158). Transnational families living in two or more countries and maintaining their relations with the homeland constitute the essential component of transnational social networks and undertake a multidimensional function in migrant lives, such as sharing products, services, assets, and economic resources.

Communication technologies play a crucial role in music sharing as well. We can examine the music-sharing practices of the Chaldean-Iraqi community in Istanbul with the homeland and third countries under the headings of circulation of recorded and live music. Young community members with permanent internet access follow the same music portals as young Assyrian-Chaldean users from different countries and share the news and latest works of Assyrian-Chaldean musicians. Circulation of live music, on the other hand, occurs when communities in different countries extend the invitation to musicians with a particular reputation among Chaldean communities for live performances at weddings, special days such as Valentine's Day, or holidays. However, the small size of the community and financial limitations prevent the Chaldean-Iraqi community in Istanbul from being a significant contributor to this type of content. This also constitutes one of the main differences between the Istanbul Iraqi-Chaldean community and other Chaldean communities worldwide, which are considered a part of the Chaldean or Iraqi-Chaldean diaspora.

The circulation of musical products among the transnational Chaldean communities around the world also shows the dynamism of the tradition. A hymn composed by a young woman, Salwa, in Iraq constitutes a striking example. Salwa composed the hymn and sent it via smartphone to her friends in Istanbul and other cities. Right after listening to the hymn on a smartphone, the Chaldean-Iraqi choir members rehearsed the hymn and performed it at the next service. Therefore, this newly composed hymn from Iraq immediately took its place in the church music repertoire in Istanbul. Besides being an instance of musical circulation among the Chaldean-Iraqi communities in various countries, this hymn's circulation also exemplifies the dynamism of religious music repertoire. Similarly, there are various other examples of newly composed hymns entering into circulation, some even based on the soundtracks of Turkish TV series.

Conclusion

Temporary migratory conditions, in general, involve a small community, constant circulation, lack of access to economic networks, and limited connections to diaspora communities. In the temporary location, there are fragile and precarious working conditions, uncertain legal status, and virtually no education possibilities. Musical practices are inevitably influenced by these conditions. While some difficulties, such as the lack or scarcity of professional musicians and the inability to carry the instruments during migration, directly affect music production, some others, like economic limitations, language barriers, limited interaction with the local communities, and the lack of institutions, have indirect consequences. Under these circumstances, musical practices in transit are quite different from those of permanent migrant communities in which second- and third-generation migrants create hybrid forms and are able to participate in an organized music market. The absence of music education or its discontinuity in the transit country also has a negative impact on the transmission of the religious and folk music repertoires for the younger generations.

Despite all these challenging factors, music emerges as one of the means that keep the community alive and vibrant, offering the community a feeling of continuity during the transit phase of their migration and helping reduce the feeling of interruption in their daily life that the migration causes. One obvious example manifests itself in the use of dance forms, which were exclusive to weddings in Iraq but became a medium of self-representation and collectivity at weekly gatherings in Istanbul. Similarly, community members are open to the support of individuals from outside the community who may be unfamiliar with their musical tradition to ensure the continuity of church music performances.

The Chaldean-Iraqi community's determination to maintain religious and nonreligious music performances persists despite the absence of professional and experienced musicians. Their insistence on expressing themselves through music and their willingness to participate in transnational music-sharing networks reveal that music is not only indispensable to community members' lives in this extraordinary phase of their migration, but it is also a means of coping with the difficulties that such an experience brings.

Notes

1 This article is derived from an earlier publication by the author published in 2016 (see references).

2 Turkey did not accept the 1967 Protocol that removes the geographical limitations of the original geographical limitation of the 1951 Refugee Convention. According to the convention, refugees from non-European countries are not accepted for settlement in Turkey.

3 According to Anthony O'Mahony (2004), a specialist on Christianity in the Middle East, there are four main Christian groups who were living in modern Iraq: 1. Catholics: Chaldeans, Syriacs (also referred to as "Syrians"), Latins, Armenians, and Greek Melkite Catholics; 2. Non-Arab Orthodox (The Church of the East, sometimes referred to as "the Assyrians"): Syriac Orthodox, Armenian Orthodox, and a small Coptic community; 3. Eastern Arab Orthodox; 4. Protestants and Anglicans who accepted Christianity in the nineteenth century. Among all these Christian communities, the biggest one is the Chaldeans, which constitutes over 70 percent of all the Christians in Iraq (O'Mahony 2004, 435) and 2 percent of the total population of the country, followed by the Syriacs and Assyrians (Hanish 2008, 32).

4 Besides these institutions, other locations, such as the houses of various community members who come together for various special events, were among my field research's prominent locations.

5 EU and Turkey signed an agreement on March 18, 2016, aimed at keeping migration flows away from Europe that includes a readmission agreement for the migrants who entered Greece illegally through Turkey.

6 Since Turkey did not remove the geographical limitation it placed on the Geneva Convention, there is no form of refugee status available for non-Europeans. In order to regulate Syrian migrants' presence in the country, Turkey issues only a "temporary protection" status for them.

7 As a dynamic concept in migration studies, diaspora has been widely discussed and redefined over the years (for some basic definitions, see: Safran 1991; Cohen 1997; Sheffer 2003; Dufoix 2008). Despite the conflicting perspectives, the concept is still in use, and I take into consideration these debates and ask whether the Chaldean-Iraqi community in Istanbul can be seen and analyzed as a diaspora community and/or what are the main differences between diasporic and transit communities in general. For further discussion, see Öğüt (2015a).

8 There is a scarcity of academic work on Chaldean music. However, when we question the concept from a musicological perspective, it is possible to state that Chaldean religious music bears similarities to Assyrian music in terms of *maqam* structure. As evinced by the Chaldeans living in northern Iraq, folk music is related to the musical traditions of other communities in the region. It is almost impossible to distinguish popular Chaldean music from the popular music of Middle Eastern countries. In this regard, the Chaldean language emerges as the most distinguishing aspect of Chaldean music.

9 The term *maqam* signifies a system of melodic practices in the Middle East and parts of North Africa, mainly Arabic, Persian, and Turkish-speaking cultures. It is based on a set of melodic patterns and their traditional use.

10 This example also reveals that assuming religious repertoire was absolute or frozen was a mistake.

11 It should be noted that the religious music repertoire is transmitted orally; the only documents about the community's repertoire are notebooks with hymn lyrics.

12 *Dabkeh* is the generic name of community line dances in the region.

13 While the inability to carry musical instruments is commonly seen in migration experiences, due to transit migration's temporary nature, possessing a new instrument at the transit location is seen as near impossible or "unnecessary." This is one of the main reasons for the lack of instrumental performances in group activities. For example, Jan could not bring the keyboard he played in Iraq to Istanbul and bought a new one to play at the church, but when he left for Australia in 2013, he left his instrument behind once again.

14 Even though musical performance's gendered structure shows parallels with the gender roles in the community, the gender aspect of both the community relations and the musical practices needs to be discussed in the broader frame. However, it is noteworthy that the migratory experience can challenge gender roles. I have witnessed some music-related cases in my research. A statement by one of my interlocutors, Jan Dark, about the effect of migration and singing in the church choir in Istanbul gives an example: "I was quite shy in Iraq and did not dare to do many of the things I do here. I came out of my shell (laughing)" (Personal interview, Jan Dark, February 4, 2013).

15 *Mawwal* is a traditional and popular vocal genre in Arabic music that is mostly free rhythm and improvisation.

16 I have witnessed a few song contests as entertainment in those events.

17 The terms Chaldean and Assyrian point to two peoples with a shared history in Mesopotamia. They speak two different dialects of Aramaic. One community belongs to the Catholic faith system, and the other belongs to the Orthodox faith system. While this distinction can be more pronounced in some destination countries, they can be perceived as one culture within the framework of shared cultural elements. Most of the sociocultural work and institutions, other than the religious ones, are referred to as "Assyrian-Chaldean." Ricky is a young Assyrian man who migrated alone and is living with the Chaldean community.

18 Iraqi-Chaldeans residing in Istanbul, to a large extent, had middle-class privileges in their homeland before migration. Didem Danış, in her report for the Center for Middle Eastern Studies (ORSAM), notes that the Iraqis who managed to arrive in Istanbul have a certain socioeconomic status. In a way, Istanbul serves the function of a filter to determine the social and economic "suitability" of migrants to become refugees. In his book *Iraqi Refugees*, Joseph Sassoon (2010), with reference to Geraldine Chatelard's work on the lives of Iraqi refugees living in Jordan during 1999–2001, notes that lower-class Iraqis do not have the same chances to leave the country or in the cases that they manage to do so, they have to sell all their property and thus end up with no economic capital to take with them.

References

Albayrak, Kadir. 1997. *Keldaniler ve Nasturiler*. Ankara: Vadi Yayınları.

Anderson, Benedict. 1994. "Exodus." *Critical Inquiry* 20 (2): 314–27.

Assmann, J. 2011. "Communicative and Cultural Memory." In *Cultural Memories: The*

Geographical Point of View, edited by Peter Meusburge, Michael Heffernan, and Edgar Wunder, 15–27. Springer: Dordrecht.

Beckerman, Michael. 2010. "Jezek, Zeisl, Améry, and the Exile in the Middle." In *Music and Displacement: Diasporas, Mobilities, and Dislocations in Europe and Beyond*, edited by Erik Levi and Florian Scheding, 43–54. Lanham, MD: Scarecrow Press.

Bohlman, Philip V. 2001. "Diaspora." *Grove Music Online*. Accessed January 18, 2021. https://doi.org/10.1093/gmo/9781561592630.article.42950.

———. 2010. "'Das Lied ist aus': The Final Resting Place Along Music's Endless Journey." In *Music and Displacement: Diasporas, Mobilities, and Dislocations in Europe and Beyond*, edited by Erik Levi and Florian Scheding, 15–29. Lanham, MD: Scarecrow Press.

Bourdieu, Pierre. 1986. "The Forms of Capital." In *Handbook of Theory and Research for the Sociology of Education*, edited by J. Richardson, 241–58. New York: Greenwood.

Cohen, R. 1997. *Global Diasporas: An Introduction*. London: UCL Press.

Conlon, Deirdre. 2007. "The Nation as Embodied Practice: Women, Migration and the Social Production of Nationhood in Ireland." PhD diss., City University of New York.

Danış, A. Didem. 2006. "Waiting on the Purgatory: Religious Networks of Iraqi Christian Transit Migrants in Istanbul." *EUI* Working Paper RSCAS No. 2006/25. https://hdl.handle.net/1814/6228.

Danış, Didem. 2007. "A Faith that Binds: Iraqi Christian Women on the Domestic Service Ladder of Istanbul." *Journal of Ethnic and Migration Studies* 33 (4): 601–15.

———. 2010. *Away from Iraq: Post 2003 Iraqi Migration to Neighboring Countries and to Turkey*. Report No. 21. Ankara: OSRAM.

Dufoix, Stéphane. 2008. *Diasporas*. Berkeley: University of California Press.

Düvell, Franck. 2006. *Crossing the Fringes of Europe: Transit Migration in the EU's Neighbourhood.* Centre on Migration, Policy and Society. Working Paper No. 33. University of Oxford.

Faist, Thomas. 2004. *The Volume and Dynamics of International Migration and Transnational Social Spaces*. Oxford: Clarendon Press.

Gallagher, B.G. 1999. "Chaldean Immigrant Women, Gender and Family." PhD diss., Wayne State University.

Hanish, Shak. 2008. "The Chaldean Assyrian Syriac People of Iraq: An Ethnic Identity Problem." *Digest of Middle East Studies* 17 (1): 32–47.

Haug, Sonja. 2008. "Migration Networks and Migration Decision-making." *Journal of Ethnic and Migration Studies* 34 (4): 585–605.

Kaiser, Tania. 2006. "Songs, Discos and Dancing in Kiryandongo, Uganda." *Journal of Ethnic and Migration Studies* 32 (2):183–202.

Kaya, Ayhan. 2002. "Aesthetics of Diaspora: Contemporary Minstrels in Turkish Berlin." *Journal of Ethnic and Migration Studies* 28: 43–62.

Landau, Carolyn. 2011. "'My Own Little Morocco at Home': A Biographical Account of Migration, Mediation and Music Consumption." In *Migrating Music*, edited by Jason Toynbee and Byron Dueck, 38–54. New York: Routledge.

Massey, Douglas S., Joaquin Arango, Grame Hugo, Ali Kouaouci, Adela Pellegrino, and J. Edward Taylor. 1993. "Theories of International Migration: A Review and Appraisal." *Population and Development Review* 19(3): 431–66.

Öğüt, Evrim Hikmet. 2015a. "Music in Transit: Musical Practices of the Chaldean-Iraqi Migrants in Istanbul." PhD diss., Istanbul Technical University.

———. 2015b. "Transit Migration: An Unnoticed Area in Ethnomusicology." *Urban People* 17 (2): 269–282.

———. 2016. "Geçici Göçün Sesleri: İstanbul'daki Keldani-Iraklı Göçmenlerin Müzik Pratikleri." *Alternatif Politika* 8 (3): 460–73.

O'Mahony, Anthony. 2004. "The Chaldean Catholic Church: The Politics of Church-State Relations in Modern Iraq." *The Heythrop Journal* 45 (4): 435–50.

Pascucci, Elisa. 2015. "Diaspora, Immobility and the Experience of Waiting: Young Iraqi Refugees in Cairo." In *Diasporas of the Modern Middle East*, edited by Anthony Gorman and Sossie Kasbarian, 338–69. Edinburgh: Edinburgh University Press.

Reyes, Adelaide. 1999. *Songs of the Caged, Songs of the Free: Music and the Vietnamese Refugee Experience.* Philadelphia: Temple University Press.

Safran, W. 1991. "Diasporas in Modern Societies: Myths of Homeland and Return." *Diaspora: A Journal of Transnational Studies* 1 (1): 83–99.

Sassoon, Joseph. 2010. *The Iraqi Refugees: The New Crisis in the Middle East.* London: IB Tauris.

Scheding, Florian. 2010. "'The Splinter in Your Eye'": Uncomfortable Legacies and German Exile Studies." In *Music and Displacement: Diasporas, Mobilities, and Dislocations in Europe and Beyond*, edited by Erik Levi and Florian Scheding, 119–34. Lanham, MD: Scarecrow Press.

Sengstock, Mary C. 2005. *Chaldeans in Michigan.* East Lansing: Michigan State University.

Shannon, Jonathan H. 2019. "From Silence into Song: Affective Horizons and Nostalgic Dwelling among Syrian Musicians in Istanbul." *Rast Musicology Journal* 7 (2): 2169–80.

Sheffer, G. 2003. *Diaspora Politics: At Home Abroad.* Cambridge: Cambridge University Press.

Solomon, Thomas. 2014. "Theorizing Diaspora, Hybridity and Music." *African Musics in Context: Institutions, Culture, Identity*, edited by Thomas Solomon, 318–59. Kampala: Fountain Publishers.

Wong, Diana. 1991. "Asylum as a Relationship of Otherness: A Study of Asylum Holders in Nuremberg, Germany." *Journal of Refugee Studies* 4 (2): 150–63.

PART FOUR

NATION-STATES AND NATIONALISM IN AND THROUGH THE MIGRATORY CONTEXT

12

Sugarcane Harvest: A Sweet, Deadly Path for Migrants in Guanacaste

David Bolaños Acuña

The sugarcane fields burst into flames every night in the northwestern province of Guanacaste, Costa Rica. As workers wake up in the morning, the sun burns incessantly from above. Only very few clouds stop its blow on the backs of the hundreds of men who complete the harvest to the rhythm of their machetes, surrounded by an endless array of ashes, heat, sweat, and sweetness.

Such a scene has been part of the everyday lives of irregular immigrants and low-paid workers who come from Nicaragua annually to find work in the fields in Guanacaste.

The laborers wrap themselves in thick coats and shirts, put on their gloves, and wear leather or rubber boots to prevent a bad cut and to hide from the scorching sun. Jugs of water and food are carried to the field in their already raw hands on a daily basis. A sip of water is taken from time to time, but the sun and the heat do not allow for real breaks, just pauses.

These men of varying ages are picked up in Liberia City in the early morning, around 3 a.m. It takes them two hours to reach the Bagatzi plantations in the town of Bagaces.

The wage for such labor is around 25 to 50 colones per meter of sugarcane cut and piled. In other words, to collect $20, each person would have to cut around half a kilometer of sugarcane daily. "There are people who don't drink water because they want to finish quickly; that's what screws them up," explained Federico, who came from Nicaragua to work that year's harvest. Guanacaste has an abnormally high rate of chronic kidney disease in the country, mostly among men of working age. Health authorities have warned about a possible link between agricultural labor and that crippling and ultimately deadly disease in the region. Since 2015, investigators in the region reached the consensus that "chronic kidney disease due to nontraditional causes has at least one risk factor: the physical effort farm workers make when they do hard jobs like cutting cane under the sun, an activity in which the illness multiplies" (Cruz 2018).

Fig. 12.1 Most of the men photographed were allowed to be photographed but preferred not to give their names for this story. Fear of retaliation by the sugarcane contractors was generalized in the field. *Photo by David Bolaños Acuña, February 2014.*

Fig. 12.2 Some men wear thick long-sleeved shirts, gloves, and hats to protect themselves from the sun, but extreme heat becomes a silent enemy under all those garments. *Photo by David Bolaños Acuña, February 2014.*

Fig. 12.3 An unnamed worker takes a brief pause to drink water. Temperatures can be over 40 degrees Celsius in Guanacaste's dry season. *Photo by David Bolaños Acuña, February 2014.*

Fig. 12.4 The fields are usually organized in rows. Workers are divided into pairs for each row, which stretch for over 500 meters. *Photo by David Bolaños Acuña, February 2014.*

Fig. 12.5 and Fig. 12.6 The completion of labor as hard as sugarcane harvesting depends on the migrant workforce. Most of the men I interviewed were Nicaraguans and lacked formal documents to be in Costa Rica. *Photo by David Bolaños Acuña, February 2014.*

Fig. 12.7 Sugarcane workers who already finished their task wait in cattle trucks for the rest of the field to be cut, their skin blackened by ashes and molasses. *Photo by David Bolaños Acuña, February 2014.*

These men do not have the luxury of complaining about their work conditions, especially when industrial machinery has been lowering the production costs for the large mills in the region. That also spares those companies from burning the fields, an essential part of the hand-cut harvest.

The only alternative for hundreds of informal workers and irregular migrants in the province to survive is to undergo what may destroy their kidneys and take their lives. "It is what it is, my brother. Where would you go instead?" said Francisco, while taking his eyes off the ground and fixing them on the row of cane he had cut thus far. It was 9 a.m., and the workday would not end until noon.

At the end of the day, the workers, all dyed dark from head to toe, climb into cattle trucks calmly, almost in resignation, to be transported back to their living quarters. Ash, sweat, and the viscous sugarcane juices form a musk that turns simple agricultural laborers into mine workers or oil extractors, all covered in a sweet black tar.

For these men, sugarcane harvest, or "zafra," is their way of earning their livelihoods but also possibly losing their lives.

References

Cruz, Maria Fernanda. 2018. "Epidemic in Guanacaste Kills Farmers Who Toil Under the Sun." The Voice of Guanacaste, 4 July 2018. Accessed April 16, 2023. https://vozdeguanacaste.com/en/an-epidemic-in-guanacaste-is-killing-farmers-who-work-long-days-under-the-sun/.

13

Representing Nicaraguan Immigration to Costa Rica through COVID-19

Carlos Sandoval-García

Introduction

Despite the fact that between 40 percent and 45 percent of total international migration takes place within countries of the Global South, most of the research and public debate focuses on South-to-North trends. This selectivity diminishes the possibilities of analyzing South-to-South migration in Africa, Asia, and Latin America. Latin American international migration, for example, is usually associated with forced populations seeking to reach the Mexican–United States border. Meanwhile, lesser attention is paid to South-to-South international migration within the region. In Central America, both intraregional and extraregional migration are a structural dimension of everyday life. It is estimated that between 12 percent and 14 percent of Central Americans live in a country different from their country of birth (Sandoval-García, 2017b). Military conflicts, economic inequalities, and, more recently, violence are among the main factors that are driving Central Americans from their countries of birth.

Similar to migration from origin countries such as Venezuela, Haiti, Bolivia, and Guatemala to destination countries such as Colombia, the Dominican Republic, Argentina, and Mexico, respectively, Nicaraguan migration to Costa Rica is a noteworthy case study of South-to-South migration in Latin America. Primarily, though not exclusively because of economic reasons, Nicaraguans have been present in Costa Rica throughout the twentieth century.

Nicaraguan immigration to Costa Rica has been a long-term process (Sandoval-García 2004). After the struggle against Somoza's dictatorship, there was migration to the United States, especially among middle and upper classes. Throughout the 1980s, migration increased in the midst of armed conflict between the Sandinista government and counterrevolutionary groups financed by the Reagan administration. Afterward, the neoliberal policies implemented by the government of Violeta Barrios from 1990, and later the devastation produced by Hurricane Mitch in 1998, in which nearly 20,000 people died in Honduras and Nicaragua, forced thousands of Nicaraguans to leave their country, leading them toward Costa Rica. In this context, Nicaraguans, once perceived as "communists," began to be represented as "migrants." This shift was a consequence of political changes in Nicaragua but also a result of the end of the socialist bloc in Europe, remembered especially for the fall of the Berlin Wall. More recently, the increase in cases of COVID-19 in 2020 was associated with Nicaraguan migrants. When the number of cases increased among Costa Rican middle and upper classes, this association lost prominence, and none took responsibility for blaming Nicaraguans for the pandemic.

According to the 2011 census, the latest carried out in Costa Rica, the migrant population in the country is about 9 percent of the total population. A total of 7.64 percent came from Nicaragua; most of them were young people between 14 and 49 years old (Bonilla and Sandoval-García 2014). There are economic sectors, mainly associated with agro-export products like pineapple, bananas, or coffee, as well as construction, private security, and paid domestic work, that are highly dependent on Nicaraguans (Sandoval-García 2017a).

Against this historical background, this article explores the forced displacement of Nicaraguans into Costa Rica, particularly as a result of the deep political crisis arising from the 2018 protests against Daniel Ortega's government. It also discusses ways in which sectors within Costa Rican society have reacted to the new arrival of Nicaraguans who, in contrast with those who came before, have a more middle-class background and whose reasons for migration are more political. This paper also seeks to analyze to what extent those who are politically motivated and recently arrived, usually known as "refugees," interact or not with those who are considered "migrants." Although they share the same national background, they appear to be divided, especially by class. Their coexistence as a minority in Costa Rica does not seem to alter the pervasive inequalities that characterize Nicaraguan society. In the context of the COVID-19 pandemic, this paper ends by reflecting on the links between the closing of borders by the Costa Rican government due to the spread of COVID-19, the refusal of the Nicaraguan government to adopt physical distancing measures, and the xenophobic response by sectors of Costa Rican society.

In methodological terms, path dependence, with a focus on historical factors, and conjunctural analysis are deployed in order to grasp the historical context of animosities between Costa Rican and Nicaraguan governments and to highlight the specific conjuncture beginning with the 2018 protests in Nicaragua through to the 2020 COVID-19 pandemic (Hall et al. 1978).

Nicaragua: From the 1979 Revolution to the 2018 Political Crisis

In 1979, the Somoza dynasty was overthrown by a popular movement led by the Sandinista National Liberation Front (FSLN in Spanish), bringing to an end the 42-year dictatorship (1937–79). A few years later, the Reagan administration began to support contra-revolutionary groups (Contras), whose support was drawn from rural areas where land expropriations faced resistance among small- and medium-sized landowners (Marchetti 2018). The FSLN was in government until 1990 when the Chamorro-led opposition won the general elections. Hyperinflation and scarcity were followed by neoliberal policies that forced the migration of thousands of Nicaraguans. During the 1980s, a large number of Nicaraguans arrived in the United States, taking advantage of Reagan's politics, according to which those leaving countries with leftist governments were authorized to stay (Rocha 2016). Years later, thousands of Nicaraguans established themselves in Costa Rica, especially in bordering regions as well as in urban areas, where more employment was available (Bonilla and Sandoval-García 2014).

In 2007, the FSLN won the general election and has been in power ever since. Daniel Ortega was elected as a result of an agreement with Arnoldo Alemán, the former president of Nicaragua (1997–2002), who was accused of corruption but later exonerated. President Ortega has concentrated a vast array of political power and economic activities together with his partner Rosario Murillo, who holds the vice presidency. What is supposed to be a government based on Christian, socialist, and solidarity principles has metamorphosed into a sort of authoritarian populist kleptocracy.

The elite private sector, namely former Sandinista adversaries, found possibilities to reproduce their businesses throughout Ortega's presidency. Buying and selling exchange rates before the 2018 crisis, for example, were higher than in the rest of the banks located in Central America (Spalding 2017).

In April 2018, resistance toward the interoceanic canal proposal using communal lands, together with the Indio Maíz Biological Reserve fire and changes to the public pension system, gave rise to massive protests, led especially by students and peasant organizations (Rocha 2019). Protests, including blockades (known in Nicaragua as "tranques"), became widespread and called for the end of government corruption and authoritarianism. Attacks on large public artworks, the so-called "trees of life" erected by Vice President Rosario Murillo, were a key symbolic action through which new generations have shown their resistance. About 320 people have lost their lives during a year and a half of protests and repression. It is too early to conclude what will be the outcome of these social mobilizations. What seems to be clear is that Nicaraguan politics will not be the same (Ramos, Baltodano, and Bellorin 2020; Ortega-Hegg et al. 2020).

The political crisis forced thousands of Nicaraguans to leave their country. During 2018 and 2019, 29,216 Nicaraguans applied for asylum in Costa Rica. However, the number of displaced people due to the crisis and persecution is difficult to estimate. Not all people who left Nicaragua have applied for asylum, and some applicants who arrived before 2018 and did not possess the appropriate documentation applied too. Among the applicants are activists, who for years resisted Ortega's policies and politics and were particularly active during the uprisings. In addition, there are those who were not always politically active but were nevertheless fleeing the violence. A refugee applicant remembers:

> My decision to leave Nicaragua was not my wish, it was only because I expressed [my views], because I was gathering groups of friends and, who were able to demonstrate to this bad government that they were killing our university and secondary school companions. We were going out with the national flag, in order to show we wanted a Free Nicaragua, but I did not wait for them to take reprisals. People who support the government began to send police officials to ask for me. On one occasion an uncle was called by phone to inform him that if I continue supporting the marches, I was going to be killed. My family decided it was better to leave the country and apply for asylum here in Costa Rica.

Toward the end of May 2018, on the anniversary of the 1979 Revolution, repression increased, and most of the *tranques* (blockades) were removed. The government of Daniel Ortega framed the protests as "terrorism," and actions were described in the past tense as if the conflict had already ceased. The government banned the use of the flag in public demonstrations because the opposition used it as a way of representing the nation. Meanwhile, the political opposition has not found an easy path to forge a broad coalition.

The Articulation of Authoritarian Populism in Costa Rica

The 2018 crisis in Nicaragua coincided with a bitterly disputed general election and later with a controversial project of tax reform in Costa Rica. The neoconservative Christian right-wing party, Restauración Nacional, won the first round with 40 percent of the vote but did not reach a majority. In the second round, the center-left party, Acción Ciudadana (PAC), won the presidency for a second term, but its representation in the national assembly (nine out of fifty-seven members) was minor. Even though Costa Rica is organized as a presidential system, an increase in the number of political parties represented in the national assembly has led it to become, in practical terms, far more like a parliamentary system in which majorities must be forged.

Meanwhile, the fiscal deficit had reached 5.57 percent of GDP, and a major tax reform proposal, largely postponed for twenty years, was sent to congress. Replacement of sales taxes with a value-added tax (VAT), increased taxes for high-income earnings, and further safeguards against tax evasion were among the initial key proposals. However, none of them were approved, since policies required for preventing evasion and elusion did not eventuate. Debate and controversy ensued about the degree to which the reforms were regressive or progressive. As PAC was in a minority position in the national assembly, and the left party, Frente Amplio, had only one parliamentary representative, the approval of more progressive policies intended to prevent evasion was hindered.

Some opposition sectors from the right and far-right argued that tax reform revenues were going to be used for social benefits for recently arrived Nicaraguans. Xenophobia was added to already existing homophobic rhetoric often used during the election campaign in social media, which portrayed PAC as a "permissive" party, supporting same-sex marriage and abortion, some of the main concerns of neoconservative ideologues. A number of far-right initiatives and profiles have been operating on Facebook, including "Alt Right Costa Rica," which resembles Steve Bannon and "Breitbart news" sites (Neiwert 2017). The homophobia and xenophobia displayed are generally based on perceived threats to Costa Rican national identity.

Under this far-right rhetoric, a march against migration took place in August 2018, which was the first occasion in which verbal violence turned into a collective action against migration in Costa Rica (Sandoval-García 2018). About six hundred people congregated in La Merced Park, known to be a common site for meetings by Nicaraguan migrants. Despite the crude reality of unemployment for thousands of Costa Ricans, demonstrators claimed that migrants lacking paid jobs should not be eligible for social benefits. Rising calls against immigration were a means of articulating discontent toward the permissive government.

The Colectivo Bienestar y Migraciones en Costa Rica (Welfare and Solidarity Network), which brings together NGOs, religious initiatives, and academics, among others, organized a countermarch that gathered around four thousand people. Peace and solidarity were key messages for calling the countermarch. The march for peace and solidarity began in Paseo Colón and the University of Costa Rica, the west and east of San José downtown, respectively. The media, usually not very friendly toward Nicaraguan immigration, supported the countermarch, which received an unusual amount of attention.

The countermarch confirmed the importance of increasing the visibility of those supporting migrants and human rights, a necessary task that returned when a small political party, self-described as "center-right," proposed a change in the Costa Rican Constitution according to which the principle of *ius solis* would be replaced by the *ius sangius* doctrine (Mena 2019). *Ius solis* states that a newborn holds the nationality of the country in which it was born, notwithstanding the nationality of their parents. *Ius sanguis* states that children inherit their parent's nationality, even in circumstances where they do not live in their country of birth. This is arguably the most radical constitutional proposal regarding migration that Costa Rica has seen. It resembles developments in the Dominican Republic, where a 2013 Constitutional Court decision established that children born in the Dominican Republic, whose parents were from Haiti, did not have the right to be Dominican citizens. This decision considerably increased the number of stateless children in the Dominican Republic (CIDH 2015).

Although difficult to quantify, those migrants who were forced to leave Nicaragua after 2018 because of political persecution appear not to have invested much effort into establishing connections with Nicaraguans who migrated for economic reasons during the previous four decades. Even though migration is indispensable for thousands of families in Nicaragua, even critical analyses take it for granted; migrating to Costa Rica is assumed to be a sort of "destiny." It seems to reproduce social divisions so ingrained within Nicaraguan society, according to which there is little shared common language between those with more formal educational backgrounds and "the rest." Overcoming this dichotomy is crucial for forging more inclusive political projects in Nicaragua and, further, for promoting encounters between those displaced by either economic or political reasons that are essential to the task of constructing social cohesion and common purpose. In terms of electoral politics, this is of utmost importance, especially because Nicaraguan electoral legislation does not allow voting from abroad, which some opposition groups aim to include in future electoral reforms.

Pandemic Times

During 2020, in the context of the COVID-19 pandemic, xenophobia resurfaced in Costa Rica. Claims emerged that the Nicaraguan economy was not prepared for the consequences of physical distancing measures and that Ortega's regime had not implemented appropriate social distancing measures, even promoting gatherings during Easter 2020, a time when most countries had implemented strong restrictions. The Nicaraguan Ministry of Health (MINSA) boasted that the very small number of confirmed cases had vindicated their approach to physical distancing measures. Meanwhile, Nicaraguan citizens had taken the initiative themselves and adopted various protection measures.

Common criticisms at the time underlined that without knowing the number of tests carried out by the MINSA, it was impossible to assess the impact of COVID-19. An increased number of people diagnosed with pneumonia, which shows similarities with the clinical manifestations of COVID-19, added new uncertainties to this scenario. Suspicions grew after news that several people who arrived in Cuba after visiting Nicaragua were diagnosed with COVID-19. At the end of May 2020, MINSA reported that 35 people died due to COVID-19 (Confidencial 2020). A Citizen Observatory of COVID-19, which provides independent information, reported the death of 805 people with clinical manifestations that suggested COVID-19 infections. The sharp difference between these two reports only confirmed the lack of data reliability, which is a consequence of subordinating COVID-19 policy to political interests (Huete-Pérez and Ortega-Hegg 2020).

As a doctor interviewed by the Nicaraguan prime-time television *Esta semana* said, "The pandemic was [always] going to arrive to Nicaragua. It was not the responsibility of the Nicaraguan government because it is a worldwide pandemic—but the ways in which it is handled is a government's responsibility" (Esta semana 2020). MINSA stopped the daily publication of cases and insisted that fumigation, usually employed to prevent dengue fever, would prevent transmission of COVID-19.

Ortega insisted that the increase in cases of pneumonia followed a pattern also present in previous years in Nicaragua. However, ignoring community transmission became impossible for Ortega's regime. Pressure increased from health specialists, but authorities were not able to recognize the huge risks underway. Specialists expected an inundation of hospitals and growing mortality trends (Ortega-Hegg et al. 2020). The World Health Organization (WHO) confirmed that Nicaragua was in the fourth stage of the pandemic, characterized by community transmission.

To what extent the pandemic and this horrific scenario would seriously undermine the legitimacy of the Nicaraguan government was and remains a pressing question at the time of writing. General elections are due to be held in November 2021, and the political reforms required to bring about transparency are far from being a reality, while the opposition has not found ways to forge a strong coalition. Although it is too early to conclude whether the social consequences of this pandemic will erode Ortega's legitimacy, his irresponsibility regarding COVID-19 increases uncertainty in the country and reinforces political polarization.

Meanwhile, Costa Rica's Ministry of Health adopted physical distancing and other measures, which included the lockdown of borders, to confront COVID-19. By the end of May 2020, nearly 1,000 cases and ten deaths had been reported in the country, one of the lowest known fatality rates. Long-term investment in public health facilities and programs is at the core of the strong Costa Rican response to the pandemic.

The irresponsibility of the Ortega government and the possibility of non-authorized crossings of Nicaraguans into Costa Rica reinforced fears about COVID-19 contagion. The Nicaragua–Costa Rica border is approximately 300 kilometers in length, including various nonauthorized points of entry. Threatening narratives of "pollution" spilling (Gilman 1985; Sandoval-García 2004) across the border stoked fears on social media. Again, as in the past, a discourse was promoted in which the Costa Rican national body was threatened by outsiders (Sandoval-García 2004).

According to the 2011 census, the last carried out in Costa Rica, 9 percent of the population were foreign-born, of which 7.3 percent were from Nicaragua. At the end of May, official figures reported that 212 (19.55 percent) out of 1,084 confirmed cases of COVID-19 were foreigners, which doubled the percentage provided by the census. Despite pressure from the media, the Ministry of Health did not disclose the nationality of foreigners diagnosed as having COVID-19, but there were multiple narratives and rumors that associated the increase of cases during May with migrants, especially Nicaraguans. Separating the fact that foreign populations are overrepresented among those who tested positive for COVID-19 from xenophobia was a pressing challenge throughout this phase of the pandemic.

At one point, the entrance of an estimated sixty-nine Nicaraguans into Costa Rican territory through a nonauthorized point became the news of the day in May 2020. It was later revealed that they were recruited by pineapple-growing companies, which were under pressure to collect the harvest (Quesada 2020). The Ministry of Health informed the companies contracting undocumented migrants that they would be fined, as required by the Immigration Law. As the literature on migration and exclusion has shown for a long time, those who are refused are at the same time indispensable (Stallybrass and White 1986). This paradox has shown to be even more true during a pandemic when threats around contagion increase and, simultaneously, migrant labor is essential to the economy. Addressing such interdependence is an ongoing task and will be addressed in the last section of this chapter.

Also, in May 2020, 14 foreigners contracted COVID-19 in a migrant detention center, and, as a consequence, the Ministry of Health's Facebook profile documented more than 34,000 comments, most of them emphasizing the threat posed by Nicaraguans. Costa Rican health authorities replied again that the nationality of those who contracted COVID-19, following the protocols, would not be revealed. The authorities consciously avoided any action that could be interpreted as playing the "migration card." Despite this, there were other comments suggesting that the government must even block the repatriation of Costa Ricans who were abroad and still seeking to return to the country.

In short, COVID-19 provided a set of circumstances in which xenophobic language could thrive without always being explicit. "Inferential racism," as Stuart Hall (1990) termed it, is highly persuasive because those who employ derogative terms speak on behalf of public health and are not immediately recognizable as "racists." "Inferential racism" then can create conditions for "overt racism," which is openly manifest. In the context of COVID-19, the image of a Rottweiler dog within the Costa Rican flag has reemerged, which remembers the death of Nicaraguan Natividad Canda toward the end of 2005, who was ferociously attacked by several Rottweiler dogs in plain view of a group of policemen and civilians (Sandoval-García 2011). Again, in May 2020, images circulating in social media suggested that Rottweilers were to patrol the Nicaragua–Costa Rica border.

In addition to the inferential and overt racism, an increased amount of fake news went into high circulation, especially through WhatsApp. There was a conscious intention to generate fear and hate (Salazar 2020; Sánchez 2020).

While the entrance of foreigners to the national territory is a permanent concern, little attention has been paid to the deaths of Costa Ricans in the

United States due to COVID-19 infections. Although Costa Rican emigration is about 3 percent of the total population and increases in recent decades have been much lower than other Central American countries, by the end of May 2020, 24 Costa Ricans had lost their lives in the United States during the pandemic. A significant number of Costa Ricans in the US live in the states of New Jersey and New York, two states where the COVID-19 pandemic was especially severe. In contrast, in the same period, just ten people died of COVID-19 in Costa Rica.

The deaths of Costa Ricans received some media attention coverage, but there was little narrative about the lives lost. Hence, it appears that the weak Costa Rican public sphere (Sandoval-García 2008) was not prepared to consider Costa Ricans as "Latinos" or "migrants" in the US context. If Costa Ricans were recognized as "others" in the US context, it may contribute to a broader reflection on how Nicaraguans are represented in Costa Rica. Additionally, a major challenge for solidarity and hospitality in Costa Rica will be to focus on the risks associated with the absence of effective public health interventions in Nicaragua and even in the United States, where the first people who tested positive for COVID-19 came from, while at the same time avoiding language that terms outsiders "pollutants" of the national body.

Throughout the period 2018–20, increases in xenophobia did not go without contestation. An account of this conjuncture would be incomplete if solidarity developments were left ignored. Xenophobia has become a word increasingly familiar beyond academic circles. Hence, there is pressure to avoid making comments that might be perceived as xenophobic. In turn, hostile voices toward Nicaraguans try to distinguish themselves from subscribing to xenophobia. The statement "I am not xenophobic, but . . ." (Sandoval-García 2020) seems to confirm that the critique of hostility and racism has been heard among those who comment on migration in social networks and everyday life.

During the pandemic, forging solidarities faced numerous difficulties. Distinguishing between genuine concerns about community transmission risks, especially due to Ortega's irresponsibility and xenophobia, was a challenge. A notable exception was a campaign produced by the Costa Rican Ministry of Justice and Peace, in which xenophobia was represented as a virus that has to be prevented, in addition to COVID-19. This offered a narrative opportunity to place xenophobia in a recognizable and important context, which might encourage resistance to hate.

Despite these efforts, hate is more prevalent than respect in social networks. Although COVID-19 does not recognize borders, its representations are highly nationalized. In the final section of this chapter, interdependence is suggested as a key missing concept for addressing the relationships between states and societies.

Interdependence: A Missing Concept

Despite there being multiple experiences of interdependence, there are few powerful narratives able to translate lived experience into the public sphere. This section explores ways in which evidence might be translated into narratives.

Although Nicaraguan migrants are vital to most sectors of the old and new export-based economy for which Costa Rica is known abroad, including coffee, bananas, and pineapple, to mention just a few, many who work in the construction sector, paid domestic work, not to mention the infrastructure sector, which has made the tourism boom possible (almost 3 million people visited Costa Rica in 2018, a country where 5 million live), find it hard to make their voices heard in the public sphere. Nicaraguans contribute about 11 percent of Costa Rica's GDP (OECD/OIT 2018). Because Nicaraguans perceive themselves as unwanted foreigners, often seen through xenophobic discourses, they do not feel authorized to speak out about their rights or their contributions, economic and otherwise, to Costa Rican society. Meanwhile, remittances sent by Nicaraguans abroad (including those who live in the United States) are estimated to be 14 percent of Nicaragua's GDP (Orozco 2020). Ironically, misrecognition of Nicaraguans abroad is also present within Nicaraguan society itself, even among the progressive left. Luis Carrión's reflection on the fortieth anniversary of the Sandinista Revolution, for example, did not include even a word about the thousands of Nicaraguans who have been forced to migrate (Carrión 2019).

Overall, narratives of migration are framed in terms of costs, while contributions are hardly addressed. Usually, the Costa Rican private sector does not recognize that Nicaraguans are vital to key economic sectors. One notable exception was after the August 2018 march against migration when a large construction company hung a banner in which the Nicaraguan and Costa Rican flags appeared together with the claim "we are a family." The CEO declared to the media: "When we saw that [the march against migrants] happening, we realized: 'One moment, they belong to our company, [xenophobia] refers to us, something has to be done. From there, we devised that message to make clear that we stand with them and that we are all one family'" (quoted in Herrera 2019).

Since the private sector hardly recognizes the degree to which interdependence characterizes the labor market, this intervention is a promising development, even if still an exception. It also confirms that the private sector could make a huge contribution in terms of mere recognition of the situation. Unfortunately, from both sides of the border, interdependence is usually underrecognized.

The pandemic reaffirmed ways in which interdependency is at the center of agro-exports and other economic activities. Borders may have been closed, but migrants were still required, though companies hardly openly recognized that they depend on a migrant labor force. Addressing this interdependency demands that we overcome exclusively "national"-based approaches. Sadly, the pandemic has exposed the lack of regional and long-term strategies within Central America to tackle the structural exclusion at the center of health systems in the region.

The experiences faced by mixed families and first-generation Costa Ricans whose parents are Nicaraguans is a second example of interdependency. A survey conducted by the National Institute of Statistics and Census of Costa Rica (INEC) concluded that 17 percent of births that took place in Costa Rica have either a Nicaraguan father or mother. Meanwhile, only 3.98 percent of births that occurred in 2018 have both parents from Nicaragua. Therefore, binational families are now a constitutive reality of contemporary Costa Rican society, and an important number of Costa Rican children have a Nicaraguan parent. However, this diversity of family backgrounds and this important first generation of Costa Ricans of Nicaraguan background does not translate into recognition, either in Nicaragua or Costa Rica.

Nicaraguan immigration to Costa Rica coincides with a long-term demographic transition that is being experienced by Costa Rican society. Costa Rica reports both the lowest Latin American fertility rate and an increase in life expectancy. The fertility rate decreased from 1.84 in 2012 to 1.67 in 2017 (Ávalos 2017). The proportion of elderly people will reach 20 percent of the total population by 2040, a transition that, in the case of Mediterranean countries, took much longer. Despite the long-term repercussions of this transition in terms of jobs, pensions, and health provision, there is little recognition of it beyond specialists.

Despite the fact that Costa Rican and Nicaraguan societies are deeply interdependent not only in history and territory but also in terms of demographic, economic, and cultural processes, there are more narratives reproducing xenophobia than those that create and reproduce notions of diversity and mutual recognition. Thus, the crucial task remains of translating interdependence into narratives (including what is taught in formal education), cultural forms, and platforms, such as literature, visual arts, and digital media. This process of translation not only depends on content, forms, and platforms, which are indispensable, but also on new generations of Costa Ricans who are prepared to share their intercultural experiences. Interdependency requires cultural forms, narratives, and digital platforms to be more visible. But what is likely even more important are people willing to speak out based on their lived experience of interdependence.

Those who are forced to leave Nicaragua share with those who have little opportunity in the host society experiences of exclusion. Instead of nation-based narratives according to which they belong to different countries and therefore have to be treated differently, it might be argued that they face inequalities, poverty, and lack of opportunities. The challenge, of course, is to address these shared experiences beyond nationalist rhetoric. Intersectionality of social dimensions such as class, nationality, race, and gender, to mention only a few, might offer opportunities to overcome narratives of "divide and conquer," according to which "foreigners and locals" are inextricably different.

In conclusion, despite the fact that economic, demographic, and social analyses provide clues to think (and, of course, feel) about interdependency, the necessary narratives are not prevalent. Experience has to be translated into narratives as part of a pedagogical and political process. Factual knowledge, creativity, and digital skills are today indispensable to achieving these goals.

To Be Continued . . .

My initial aim when I began research for this paper was to analyze the political crisis in Nicaragua that resulted in the death of at least 328 people and the forced displacement of Nicaraguans to Costa Rica. However, this changed when COVID-19 upended the world, and the Nicaraguan government resisted widely accepted health measures such as physical distancing. By the beginning of June 2020, nearly 1,000 people (in a country of 6.4 million population) might have died due to COVID-19 in Nicaragua, and it is expected that this number will have increased considerably.

During the first three months of the pandemic, public debate was dominated by the risks of nonauthorized crossings of Nicaraguans who might be COVID-19 positive. A few weeks later, in June, the reality was characterized by the fast spread of COVID-19 throughout Nicaraguan territory.

The Costa Rican government did not make information public on how many Nicaraguans who tested positive for COVID-19 recently entered the country or had already been residing in Costa Rica before. At least some of them were working in agro-export companies in the northern and Caribbean regions of the country, and owners seem not to have stopped activities in the processing plants. This contributed to the spread of the virus. Ironically, private owners and the Ortega regime, who are supposedly located in different positions in the political spectrum, coincided in their irresponsible approaches to the pandemic.

Toward the end of 2020, almost without notice, the percentage of foreign people who were positive began to show a consistent decrease. In July 2020, about 30 percent of cases of COVID-19 in Costa Rica were foreigners, especially Nicaraguans. The high spread within processing plants located in the northern region continued in urban areas, especially in impoverished residential areas where a high number of people share small living spaces. More cases in urban areas reinforced inferential and overt racism. In December 2020, foreign people who had tested positive represented only 13 percent of the total cases (169,321), a percentage near to the estimation of the foreign population living in Costa Rica. Transmission patterns seem to have been less concentrated among impoverished communities and more spread among different social strata. For example, general social interaction, more than workplaces, were the most prevalent sites for contagion (Miranda and Carvajal 2020). Meanwhile, there does seem to be an improved understanding of the dangers of blaming COVID-19 increases on the foreign population. The peak of xenophobia appears to have passed, though it seems to be dormant in the collective unconscious.

Instead of narratives of the body politic threatened by foreign carriers of COVID-19, the hardening of the economic crisis (unemployment rose from 12 to 24 percent; poverty increased from 21 to 24 percent) took prominence. As in many other societies, far-right narratives underestimated and even pretended to ignore the complexity of the pandemic. In Costa Rica, for example, the term “PACdemic” was coined to suggest that the only crisis was the government itself, which decreed a number of lockdowns, which were interpreted as freedom restrictions.

During 2021, vaccination was the main topic of controversy. Although far-right collectives pretended to mobilize the population against vaccination, most of the population agreed to receive several doses. In fact, Costa Rica comes in sixth place in Latin America in terms of people who completed the recommended number of doses, with minor differences with countries such as Uruguay, Peru, and Argentina, whose reports are slightly higher (Statista 2022).

Nevertheless, the general election of 2022 did not favor PAC, which did not even win a place in the parliament. The new president is a former economist of the World Bank, who was accused of sexual harassment (World Bank

Administrative Tribunal 2021) and ran as an outsider who promised to solve problems that undermine confidence and social well-being in Costa Rica. His main slogan was "Yo me como la bronca" ("I'll take the heat"), which means that "I," a male individual, will be able to sort out the challenges faced by the Costa Rican society as a whole. It adds more evidence to the populist and masculinist rhetoric that is pervasive in contemporary politics in many societies.

Meanwhile, since May 2021, Ortega's regimen cancelled electoral credentials to opposition political parties and detained opposition leaders in Nicaragua, including former commanders of FSLN Víctor Hugo Tinoco, Dora María Téllez, and Hugo Torres, who died in prison. Official results of the 2021 elections gave Ortega 75.92 percent of valid votes for a third consecutive presidency term and 75 out of 90 for the parliament, which assured an absolute majority (Sandoval-García and Rodríguez 2022). FSLN won in all 153 municipalities throughout the country, which is a very rare coincidence, especially for a country in which there were no reliable official statistics on cases of COVID-19 and where measures of physical distancing were a decision of citizens under the vacuum produced by the lack of public policies.

Under these circumstances, nearly 100,000 Nicaraguans have arrived in Costa Rica since 2018 (CEJIL 2022), and the possibility for bilateral migration agendas between both governments is improbable. What is even worse is that there is no recognition of the necessity of such agendas.

References

Ávalos, Ángela. 2017. "Nacimientos en Costa Rica se reducen a menos de 7000 por año." *La Nación*, April 20, 2018. https://www.nacion.com/el-pais/salud/nacimientos-en-costa-rica-se-reducen-a-menos-de-7/THD5YRFDHZDGTMXHCJJC2LL3CU/story/.

Bonilla, Róger Carrión, and Carlos Sandoval-García. 2014. "Aspectos sociodemográficos de la migración nicaragüense en Costa Rica según el Censo 2011." *Simposio Censo 2011*, 260–80. San José: INEC.

Carrión, Luis. 2019. "Nicaragua 40 años de la revolución nicaragüenses. ¿Pudo haber sido de otra manera?" *Envío,* No. 448. Accessed June 29, 2021. https://www.envio.org.ni/articulo/5650.

CEJIL. 2022. "Observaciones a las recomendaciones emitidas por el CIDH en el Informe sobre Migración forzada de personas nicaragüenses a Costa Rica." https://cejilmovilidadenmesoamerica.org/wp-content/uploads/2022/06/Informe-Recomendaciones-Nicaragua-y-Costa-Rica-1.pdf.

CIDH (Comisión Interamericana de Derechos Humanos). 2015. *Situación de los Derechos Humanos en República Dominicana.* Washington: Organización de Estados Americanos.

Confidencial. 2020. "Minsa admite 759 casos y 35 fallecidos por covid-19 en Nicaragua." Accessed June 29, 2021. https://confidencial.digital/nacion/minsa-admite-759-casos-y-35-fallecidos-por-covid-19-en-nicaragua/.

Esta semana. 2020. “El brote de covid-19 en Chinandega, los hospitales abarrotados, y el régimen calla.” *Confidencial*, Youtube video, May 10. Accessed June 29, 2021. https://www.youtube.com/watch?v=ZlQYIJs46c4.

Gilman, Sander 1985. *Difference and Pathology: Stereotypes of Sexuality, Race and Madness*. Ithaca: Cornell University Press.

Hall, Stuart. 1990. “The Whites in Their Eyes: Racist Ideologies and the Media.” In *The Media Reader*, edited by Manuel Alvarado and John Thompson, 8–23. London: British Film Institute.

Hall, Stuart, Chas Critcher, Tony Jefferson, John Clarke, and Brian Roberts. 1978. *Policing the Crisis: Mugging, the State, and Law and Order.* London: Macmillan.

Herrera, Juan José. 2019. “Peones extranjeros sostienen más del 70% de la agricultura y del 50% de la construcción tica.” Teletica, June 7, 2019. Accessed June 29, 2021. https://www.teletica.com/227617_peones-extranjeros-sostienen-mas-del-70-de-la-agricultura-y-del-50-de-la-construccion-tica.

Huete-Pérez, Jorge, and Manuel Ortega-Hegg, eds. 2020. *Covid-19, el caso de Nicaragua. Aportes para enfrentar la pandemia*. Managua: Academia de Ciencias de Nicaragua.

Marchetti, Pedro. 2018. “Ricardo Falla. El detalle del tríptico de su vida: formación, investigación e incidencia.” In *Eutopía*, Entrega especial: 133–203.

Mena, Sergio. 2019. “Proyecto para reformar Ley de Migración y Extranjería.” San José: Asamblea Legislativa.

Miranda, Hulda, and Rigoberto Carvajal. 2020. “Así fueron los clústers de covid-19 en Costa Rica hasta el mes en que se desbordó la pandemia.” *Interferencia*, September 16, 2020. Accessed June 29, 2021. https://radios.ucr.ac.cr/2020/09/interferencia/asi-fueron-los-clusters-de-covid-19-en-costa-rica-hasta-el-mes-en-que-se-desbordo-la-pandemia/.

Neiwert, David. 2017. *Alt-America: The Rise of the Radical Right in the Age of Trump*. London: Verso.

OECD/OIT. 2018. “Cómo los inmigrantes contribuyen a la economía de Costa Rica.” *OECD Library*, July 30, 2018. Paris: Éditions OECD. Accessed June 29, 2021. http://dx.doi.org/10.1787/9789264303867-es.

Orozco, Manuel. 2020. “La migración y la democracia: exclusión y expulsión.” Confidencial. Accessed June 29, 2021. https://confidencial.digital/opinion/la-migracion-y-la-democracia-exclusion-y-expulsion/.

Ortega-Hegg, Irene Manuel, Agudelo Builes, Jessica Martínez Cruz, Mario Sánchez, Hloreley Osorio Mercado, Jessica Pérez Reynosa, Sergio Ramírez, Hellen Castillo Rodríguez, and Juan Pablo Gómez, eds. 2020. *Nicaragua 2020. La insurrección cívica de abril 2018*. Managua: UCA.

Quesada, Gerardo. 2020. “Contratistas con complicidad de empresas, estarían detrás de la explotación de indocumentados en actividad agrícola.” El Norte Hoy, May 8, 2020. Accessed June 29, 2021. https://elnortehoy.wordpress.com/2020/05/08/contratistas-con-complicidad-de-empresas-estarian-detras-explotacion-de-indocumentados-en-empresas-agricolas/.

Ramos, Alberto, Umanzor López-Baltodano, and Ludwig Moncada Bellorin, eds. 2020. *Anhelos de un nuevo horizonte. Aportes para una Nicaragua democrática*. San José: FLACSO.

Rocha, José Luis. 2016. "Evolución de la ilegalidad migratoria de los centroamericanos vista desde un censo, la geopolítica y los modelos migratorios." In *Migraciones en América Central: Territorios, políticas y actores*, edited by Carlos Sandoval-García, 119–38. San José: Editorial Universidad de Costa Rica.

———. 2019. *Autoconvocados y conectados: los universitarios en la revuelta de abril en Nicaragua.* San Salvador: UCA Editores y Fondo Editorial UCA Publicaciones.

Salazar, Daniel M. 2020. "Video sobre entierros colectivos no se grabó en Nicaragua, sino en Ecuador." *Radios UCR.* April 30, 2020. Accessed June 29, 2021. https://doblecheck.cr/video-sobre-entierros-colectivos-no-se-grabo-en-nicaragua-sino-en-ecuador/.

Sánchez, Alexander. 2020. "Video sobre decenas de féretros apiñados en un cementerio no son de Nicaragua." *La Nacion.* April 12, 2020. Accessed June 29, 2021. https://www.nacion.com/no-coma-cuento/nocomacuento-video-sobre-decenas-de-feretros/KW4LYUYDBRC7RKIP5JMLKLGADE/story/.

Sandoval-García, Carlos. 2004. *Threatening Others: Nicaraguans and the Formation of National Identities in Costa Rica.* Athens: Ohio University Press.

———. 2008. "Costa Rica: Many Channels, Scarce Communication." In *The Media in Latin America*, edited by Jairo Lugo, 110–15. Buckingham: Open University Press.

———. 2011. "Introduction." In *Shattering Myths on Immigration and Emigration in Costa Rica*, edited by Carlos Sandoval, xvii–xxvi. Maryland: Lexington Books.

———. 2017a. "Nicaraguan Immigration to Costa Rica: Tendencies, Policies, and Politics." *Global Latinos: Latin American Diasporas and Regional Migrations*, edited by Mark Overmyer-Velázquez and Enrique Sepúlveda, 95–114. Oxford: Oxford University Press.

———. 2017b. *No More Walls: Exclusion and Forced Migration in Central America.* Cham, Switzerland: Springer, Palgrave Pivot.

———. 2018. "Tarjeta roja a la xenophobia." In *Semanario Universidad.* August 24, 2018. Accessed June 29, 2021. https://semanariouniversidad.com/opinion/tarjeta-roja-a-la-xenofobia/.

———. 2020. "¿Por qué la migración se convierte en la bandera de las derechas? Consideraciones para una política progresista." In *Puentes, no muros. Contribuciones para una política progresista en migraciones*, edited by Carlos Sandoval-García, 121–42. Buenos Aires: CLACSO.

Sandoval-García, Carlos, and Bryan Rodríguez. 2022. "Nicaragua: Lo que revela la base de datos empotrada del CSE sobre las elecciones de 2021." Sin Permiso. 20 October. https://www.sinpermiso.info/textos/nicaragua-lo-que-revela-la-base-de-datos-empotrada-del-cse-sobre-las-elecciones-de-2021.

Spalding, Rose J. 2017. "Los empresarios y el Estado posrevolucionario: El reordenamiento de las élites y la nueva estrategia de colaboración en Nicaragua." *Anuario de Estudios Centroamericanos* 43: 149–88. Accessed June 29, 2021. https://revistas.ucr.ac.cr/index.php/anuario/article/view/31556.

Stallybrass, Peter, and Allon White. 1986. *The Politics and Poetics of Transgression.* Ithaca: Cornell University Press.

Statista. 2022. Porcentaje de vacunados y dosis administradas contra el coronavirus (COVID-19) en América Latina y el Caribe a 14 de junio de 2022, por país. Accessed July 20, 2022. https://es.statista.com/estadisticas/1258801/porcentaje-y-numero-vacunados-contra-covid-19-en-latinoamerica-por-pais/.

World Bank Administrative Tribunal. 2021. World Bank Administrative Tribunal Annual Report. https://tribunal.worldbank.org/sites/default/files/documents/FY-2021-World-Bank-Administrative-Tribunal-Annual-Report.pdf.

14

Displaced Venezuelan Migrants in Colombia: An Example of Global Oppression

Allison B. Wolf

> *We aren't here because we want to be. We're here because the situation forced us to leave our families and everything behind.*
>
> —Grebelin Muñoz, 27 (Rojas and Carpentiero 2020)

> "There are bad people in Colombia but that's the same everywhere. . . . Sometimes people insult you . . . but nothing too bad has happened to me personally. Sometimes people give us food even if we don't ask for it."
>
> Gabriel Lugo, 32 (Stott and Long 2020)

In 2020, the number of Venezuelan nationals who fled their homeland was estimated at over 5.2 million and was expected to surpass 7 million in 2021, making it the world's largest exodus (Organization of American States 2020). Almost 30 percent of these displaced migrants go to Colombia, making the South American nation the primary recipient of Venezuelan nationals in Latin America (IOM 2020). As of December 2022, the Colombian migration authority, Migración Colombia, approximated that 2.48 million Venezuelan nationals were living in Colombia (Plataforma de Coordinación Interagencial para Refugiados y Migrantes de Venezuela 2022). Despite the COVID-19 pandemic and the Colombian government's response to it, which forced tens of thousands to return to Venezuela (Pozzebon 2020; Otis 2020), the population of displaced Venezuelans in Colombia remained high and was expected to continue rising in 2021 (Infobae 2020).

This situation was a relatively sudden reversal of circumstance. Before the Colombian government signed a peace accord with the Revolutionary Armed Forces of Colombia (FARC) in 2016 to end the internal armed conflict, Colombia was largely a country of emigration, with many of its citizens going to Venezuela.[1] Originally drawn by the allure of better economic opportunities, living conditions, and public services in the 1950s, Colombian migration really began in earnest in the 1970s as millions fled the internal armed conflict between the government and a number of left-wing insurgent groups, FARC, the Popular Liberation Army (EPL), the National Liberation Army (ELN), and the 19th of April Movement (M-19) (*El Tiempo* 2018). While official records were not kept, the director of the Consultancy for Human Rights and Displacement (CODHES), Marco Romero Silva, estimates that millions of Colombians went to Venezuela during the 1980s and 1990s, and Universidad Central de Venezuela professor Antonio de Lisio puts the numbers as high as 5 million (LaRosa and Mejía 2017, 215). Even in the mid-2000s, roughly 1.8 million Colombians still resided in Venezuela (Janetsky 2019). The tide began changing once Hugo Chavez assumed power and wealthy, highly educated Venezuelans left for the United States, Spain, and Colombia (LaRosa and Mejía 2017, 221). They were followed by middle-class Venezuelan professionals (*El Tiempo* 2019). In addition, starting around 2015, Venezuelans began to flee their country amid a complex humanitarian crisis with an economy in free fall, runaway inflation, a collapsed healthcare system, pervasive political instability and unrest that left many without food or access to medical treatment (IACHR 2019, 513–15).

At about the same time, the relationship between Colombia and Venezuela around immigration policy began changing. In August 2015, Nicolás Maduro faced a drone attack that he initially blamed on then-Colombian president Juan Manuel Santos, who, a few days before, had granted 440,000 Venezuelans permission to stay in Colombia (LaRosa and Mejía 2017, 222). In response, Maduro shut the border and expelled thousands of Colombians living in the frontier area (Castillo Arenas and Ammerman 2020). In August 2016, Santos and Maduro met in Venezuela and agreed to reopen the border with five monitored checkpoints (LaRosa and Mejía 2017, 222). When the border reopened, the current wave of Venezuelan immigration began, with tens of thousands crossing into Colombia on the first Sunday alone (Baddour 2019a). After ninety days, the numbers reached 65,000. After a year, 470,000 displaced Venezuelans had gone to Colombia (Baddour 2019b). By November 2018, the number grew to 1 million, and over 1.8 million displaced Venezuelans were estimated to be living in

Colombia by the end of February 2020 (IOM 2020). It is approximated that fewer than 40 percent of these migrants have high school education, and the vast majority work in the informal economy (*El Tiempo* 2020a). This essay offers a general overview of their situation and suggests that it is fundamentally permeated by global oppression.

General Overview of Colombia's Response to Recent Venezuelan Migration

> "For those who want to make xenophobia a political path, we adopt the path of brotherhood . . . For those who want to outcast or discriminate against migrants, we stand up today . . . to say that we are going to take them in and we are going to support them during difficult times."
>
> Ivan Duque, President of Colombia

> "Venezuela helped many Colombians . . . Now it's Colombia's turn to help Venezuelans."
>
> Rossana Tua, 33, resident of La Magdalena, where she's lived for almost a year since moving from Venezuela

Colombia was largely unprepared for the influx of displaced Venezuelans just described and has largely been slow to respond despite public pronouncements like those just cited. In fact, Colombia did not even attempt to estimate the numbers of displaced Venezuelans in their territory until 2018 (Louidor et al. 2019). For example, Colombia neither altered its legal framework in response to the Venezuelan situation nor developed a national strategy to address displaced Venezuelans' needs, instead relying on ad hoc measures related to getting the population into a regular status, protecting their human rights, and getting them access to healthcare, education, and employment (Sanabria and Toro 2019). This should not imply, however, that Colombia made no effort. On the contrary, unlike many nations in the region, Colombia has tried to welcome its neighbors. In fact, deputy director for the International Rescue Committee Trisha Bury said: "I've never seen a government trying this hard to register people and leave the borders open" (Baddour 2019a). With the exception of special restrictions during the COVID-19 crisis, Colombia has kept its borders open for Venezuelans, extended Colombian citizenship to children born to Venezuelan parents on Colombian soil (it does not grant birthright citizenship) (Bartsfield 2019; Kurmanaev and González 2019), created a special temporary permit (PEP) allowing Venezuelans who entered Colombia with documents to work and access education and healthcare, and is working with local authorities to help Venezuelans receive better access to humanitarian assistance and become socioeconomically integrated into Colombian society (Ajiaco 2020). Beyond this, many high-ranking government officials try to set a welcoming tone and condemn anti-Venezuelan violence and xenophobia.

These efforts are insufficient because "the scale of this crisis, and the speed at which it changes, is more than Colombia can handle" (Ajiaco 2020). As the director of the Wilson Center's Latin American Program, Cynthia Arnson, echoes, regardless of its good intentions, Colombia simply does not have the "financial wherewithal to provide shelter, food, medical care, and employment to such large numbers of hungry and vulnerable people. Public health and education are already overextended and under-resourced . . . and recent migrants are sicker than in the past" (Arnson 2019). For example, the city of Bogotá announced an investment of around 50 million dollars to help Venezuelan migrants in the city access education, healthcare, and other services. These funds have provided essential information to around 11,000 Venezuelan migrants, given thousands access to healthcare, helped 2,800 children, and paid for around 26,000 Venezuelan children to enroll in school (Cortés 2019). However, as of the time of writing this chapter, in 2019, Migración Colombia estimates that there are 400,000 migrants in the city (Alcaldía de Bogotá 2019). So, even the good programs just highlighted (which helped 40,000–50,000 migrants) still leave many in need of support. This was the case *before* COVID-19 started ravaging the globe and wreaking further havoc on Colombia's already fragile healthcare system and economy. In other words, even if Colombia wanted to completely meet all of the needs of displaced Venezuelans in its territory, it simply cannot.

This is often not for lack of trying. Colombia and international nonprofits have pleaded with the international community for help to no avail. For example, even though the World Bank estimated that Colombia spent "roughly $900 million [in 2018] to meet only the basic needs of Venezuelan migrants, . . . [its] 2019 campaign . . . to help raise funds to assist Colombia in settling Venezuelan migrants raised only $32 million" (Kurmanaev and González 2019). "The 2020 appeal called for over $782 million to assist Venezuelan refugees and migrants, but, as of December 1, only 37.5% had been funded" (Welsh 2020a). Additionally, in 2020, the United Nations released a report noting that Colombia will need $641 million to meet the needs of this population in 2021, but little widespread response was forthcoming at the time of writing (early 2021). While the international community spends about $2,000 per Syrian refugee, it spends only $180 per displaced Venezuelan (Infobae 2020).

Although many have offered explanations for this poor showing—the world's resources are depleted, this is considered a political crisis of Venezuela's own making, the international community may be waiting on the sidelines to contribute to Venezuela's reconstruction after Maduro falls, there is "compassion fatigue" in Europe and the United States—they seem unconvincing given that, as Arnson again notes, "compassion fatigue hasn't stopped the international community from providing more than $17 billion[2] in assistance for Syrian refugees in less than a decade" (Arnson 2019). It appears the world is failing to meet this challenge, and Colombia cannot meet the needs of displaced Venezuelans despite its efforts.

Venezuelans in Colombia: Experiencing Global Oppression Every Day

> "I would like to stay in Bogotá; you come with the goal of helping your family, but not under these circumstances. There are no jobs. There's nothing. You can only get what you need to survive so the situation is hard."
>
> Dania Ortiz, 50 (Rojas and Carpentiero 2020)

Most scholars analyze the experiences of displaced Venezuelans (and why they are unjust) via the framework of human rights (after all, the Venezuelan government is violating its citizens' human rights—thus causing them to need to flee—and displaced Venezuelans' human rights are not being protected in Colombia).[3] I do not think this is the best approach; the human rights framework fails to provide the most accurate or complete picture of life in Colombia for displaced Venezuelans or how it is characterized by injustice, largely because it tends to overlook structural phenomena (Wolf 2019). Instead, I think that it is more apt to explore the experience via a framework of global oppression. In other words, while in legal circles, for example, immigration justice is fundamentally seen as being about protecting immigrants' human rights, I define immigration (in)justice in relation to global oppression (Wolf 2020). Put differently, we should understand that immigration justice is fundamentally about identifying and resisting global oppression against immigrants and others affected by immigration policies. As such, evaluating the extent to which policies, practices, and norms related to immigration are just requires asking how they do (or do not) create, perpetuate, and/or reflect global oppression (Wolf 2021, 2020, 2019). When we do this, we will see that a ubiquitous feature of the experience of displaced Venezuelans is that it is a constant confrontation with different aspects of global oppression. To support this point, I will briefly define "global oppression" and then show how it permeates the daily lives of displaced Venezuelans in Colombia.

Global Oppression: A Brief Overview

The basic idea behind global oppression is that it is a systemic and structural phenomenon that molds and impedes the oppressed in ways that cause them to be victims of structural injustice (Frye 1983; Wolf 2020). It occurs when a nation's policies and practices come together to place nations, transnational communities, and their members into double binds that ensure they have no good options and are vulnerable to negative consequences regardless of what they do in the specific circumstance. And they are placed in these double binds simply because they are members of those nations or transnational communities.

Under global oppression, members of specific nations are trapped in a metaphorical (and with immigration, sometimes a literal) cage by another nation's policies, practices, and norms affecting other countries. As a result, it is not accidental, random, or avoidable, but it results from the accepted policies and practices of a nation. Given the diversity of these policies and people's social and national identities, however, global oppression is not expressed in a universal way. On the contrary, it manifests in at least six respects, including exploitation, xenophobia, and systemic violence. And all of these represent different aspects of injustice Venezuelan immigrants endure every day.

The Cage of Documentation (Or Lack Thereof)

> "I came with an expired passport, my Venezuelan identity card and a card giving me permission to migrate and cross the border [which is obtained online and paid for in Venezuela] . . . They have told me to get PEP but I do not know where to get that."
>
> Dania Ortiz, 50 (Rojas and Carpentiero 2020)

> "Evidently, there are only jobs in the informal economy. Why not look for something better you may ask? Because if you don't have papers, you are useless. Or you're only useful for certain things."
>
> Paveli (Pulgarín 2017)

"I have heard so many times when I go to look for a job, 'it's a shame but I cannot interview you because you do not have a Colombian cédula.' . . . The temp agency told me they wanted to hire me but that they could not because the resident visa that I have did not meet the requirements to work in the country. . . . They even sent me to *Migración* to get a letter certifying that I was eligible to work . . . I think all of this happens because people are ignorant."

Andrea (Pulgarín 2017)

At a general level, Colombian immigration policies around documentation come together to trap these migrants in a metaphorical cage from which it is difficult to escape. Over half of displaced Venezuelans—between 790,000 and 850,000—are undocumented (Sanabria and Toro 2019; Louidor et al. 2019). Worse, it is nearly impossible for these migrants to obtain the documents required to access their entitlements since (1) there are no guidelines for how documents (like high school diplomas) can be recognized in Colombia; (2) most people cannot afford to get the documents; (3) it is almost impossible to get certified documents from Venezuela; and, related, (4) the Venezuelan government refuses to issue passports, with some waiting for months or even years (Zuniga 2018). The problem is that displaced Venezuelans can only access certain rights and services under the Colombian Constitution—such as PEP, the public education system, and nonemergency healthcare—if they have the very documents that are nearly impossible to obtain. So, they are trapped in a double bind—stay and wait for a passport or other documents (thus continuing to expose themselves to violence, starvation, and lack of medical care) or come without them and be marginalized and frozen out of opportunities in Colombia, and be vulnerable again to poverty, violence, and illness. The documentation requirements, in other words, make it such that Venezuelans must choose between the impossible—suffer at home or suffer global oppression in Colombia in at least three ways—exploitation, xenophobia, and systemic violence.

Exploitation

"We don't get a contract and they don't pay for health insurance. They pay us by the day without any receipt . . . I understand that there are many undocumented people without passports, without PEP, that just arrived but I have been here for five years. I have a passport, PEP, and all my documents."

Jessica Aponte (Ramírez and Gómez 2020)

> "People have the mistaken idea that Colombia is like the Promised Land, but it's not. Here many Venezuelans are living a hard life. They pay undocumented workers 20000 pesos per day [around $6–$7] . . . and make them work up to 12 hours per day every day of the week."
>
> José Daniel Luzardo, 23 (*Maduradas* 2018)

In *Justice and the Politics of Difference*, Iris Marion Young notes that there are multiple manifestations, or faces, of oppression. One is exploitation, or the systemic transfer of the labor from one social group to benefit another (Young 1990). When oppression manifests as exploitation, one group does most of the work but receives little to no benefit from their labor; the exploited must work in dangerous conditions for low wages and without job stability, security, or opportunity for advancement. Other parties benefit from the labor of the exploited, though, in the form of high profits, little accountability, and lower prices. In this way, the labor of one group (the exploited) is transferred to benefit other groups. In the particular context of global oppression, exploitation refers to the structurally supported transfer of the labor of members of one nation, national government, or transnational community to benefit another. I argue that part of the daily experience of displaced Venezuelans involves enduring double binds that ensure their labor will be transferred to benefit Colombians.

For the reasons already outlined, most Venezuelans are forced to work in the informal economy; they work in factories, agriculture, construction, door-to-door delivery, domestic labor, sex work, selling incidentals on street corners, begging, or receiving rent in exchange for work in the same dwelling (*El Tiempo* 2019). One study showed that only 25 percent have formal labor contracts, while another put the number even lower, at 10 percent (*El Tiempo* 2020a). Unsurprisingly, Venezuelan workers are paid less than Colombians for the same job, with the Colombian Ministry of the Treasury reporting that the wage gap is 34.9 percent (*El Tiempo* 2020a). Additionally, Venezuelans often work longer hours and in worse conditions than Colombians, with a full-time Colombian employee working 44 hours per week and most Venezuelans working 50 hours per week or more (*El Tiempo* 2020a). One local official, Juan Fernando Gómez, reported that there were "cases where people work between 14 and 18 hours a day without any social security benefits" (Restrepo 2018). If workers report these violations, companies often threaten them with deportation (Ramírez and Gómez 2020).

The situation is even worse for female workers. Thousands of desperate Venezuelan women, including professionals such as doctors, turn to sex work to survive because there are few other options (Zulver 2019; International Refugee Committee 2018). For those women who find other employment, for example, work at a restaurant, factory, or as a domestic employee, their positions are vulnerable—often getting fired if they become pregnant or have an accident (Ramírez and Gómez 2020). This happened, for example, to Mayra, who was cleaning machines in a food processing plant where she worked for three years when the machine she was cleaning turned on and caught her hand inside as the blades rotated. By the time two coworkers arrived to assist her, it had cut up much of her hand and completely severed her thumb (Ramírez and Gómez 2020). When her bosses were informed of what happened, they denied that the accident occurred in their factory, did not offer to take her to the hospital, and then accused her of purposefully putting her hand in the machine to "get money out of them" (Ramírez and Gómez 2020). Now, Mayra says: "I get depressed a lot because I am alone. Everyone is going to work" (Ramírez and Gómez 2020).

The above indicates that displaced Venezuelans' experiences are infused with exploitation. The benefits of their labor are clearly being systemically transferred to Colombians at every level of society. Colombian employers benefit from employees who work longer hours for less pay, Colombian families benefit from Venezuelan domestic labor, and the average Colombian citizen benefits from their willingness to perform certain services, such as cheaply delivering goods (food, liquor, toys, books, etc.) at nearly any hour of the day, for their comfort. Moreover, the Colombian government benefits because, while displaced Venezuelans perform essential tasks throughout the city, like separating trash or recycling, the government avoids paying for their healthcare and other social security benefits, like pensions. So, Colombian migration policies create situations that put displaced Venezuelans in the position of being exploited for the benefit of Colombians. Moreover, this occurs precisely *because of their status as displaced Venezuelans*; it is their national identity, economic position, and migration status that explains why they are in this position.

Xenophobia

> "It makes me really uncomfortable how they speak and call us things like veneca. . . . I couldn't work listening to stuff like that [especially knowing] that the owner accepted such things."
>
> Grebelin Muñoz, 27 (Rojas and Carpentiero 2020)

> "There's too much xenophobia here . . . The Colombians think we're all thieves and prostitutes. So, I guess I'm a stereotype. But it's not like I had a choice. If I could open a business, I would. But the only job a veneca (slang for a Venezuelan-Colombian) can get here is for less than [the] minimum wage in a shop or a restaurant."
>
> Paula (Collins 2019a)

Anti-Venezuelan xenophobia is widespread in Colombia. A December 2019 Gallup poll found that most Colombians view Venezuelan migrants as a problem, and 69 percent view them "unfavorably" (Guzmán and Ponce 2020). Displaced Venezuelans are constantly barraged with derogatory names, images, and language portraying them as dirty invaders, violent criminals, prostitutes, and spreaders of disease. It is near impossible for a Venezuelan like Aldena, a single mother of three, to avoid being called the racial/ethnic derogatory slang for Venezuelan immigrants, *veneco/veneca*, hear her boss proclaim, "Venezuelans are useless," or see "*Fuera Venezolanos*" on the walls of Bogotá neighborhoods (Grattan 2020). Social media is just as bad, filled with threatening memes and messages, like one picturing a man with a relieved expression on his face with the caption: "When you thought you ran over a dog but realized it was just a Venezuelan" (Memes Ran-Damn 2019) or the Twitter post saying: "Will pay a million Colombian pesos for a Venezuelan killed by rats" (Proyecto Migración Venezuela 2019).

Beyond the blatant threats and insults, Venezuelans are constantly accused of stealing Colombian jobs and draining their social resources. According to the Oxfam International report *Yes, But Not Here*, 70 percent of Colombians think that Venezuelans are a threat to their job prospects or salary, and 80 percent think they have caused public service systems, such as healthcare and education, to collapse (Rivero 2019). All these ideas converged in late 2020 to create prominent anti-Venezuelan campaigns. One such campaign involved the publication of a pamphlet ordering everyone to fire Venezuelan immigrants and hire Colombians within forty-eight hours (CNN Español 2019) and promoted deporting displaced Venezuelans (or worse) in order to "clean up" northern Colombian cities from the "delinquency" Venezuelans bring to the country (CNN Español 2019). These ideas are becoming more publicly accepted. For example, the Colombian city of Maicao's newly elected mayor, Mohammed Dasuki, combined homophobia, sexism, and anti-Venezuelan sentiment to accuse displaced Venezuelans of undercutting locals in the job market and bringing prostitution—"not just of women but also of homosexuals"—to his city (Stott and Long 2020). Luis Eduardo

Castro, the mayor of Yopal, in the eastern Colombian province of Casanare, started deporting Venezuelan migrants whom he accused of "breaking the law and has threatened to fine drivers who give lifts to refugees" (Stott and Long 2020). Beyond this, when Colombia erupted in protests in November 2019 with riots and looting breaking out in some places, it was common to hear that Venezuelans (not Colombians) were responsible for the violence and the government deported dozens, often without due process (González 2019; City Paper 2019).

More instances of anti-Venezuelan xenophobia were occurring at the time of writing. Some Colombians were protesting in solidarity with Venezuelans who could not work (because of quarantine restrictions, for example) because they lacked housing or were evicted during the COVID-19 pandemic (Gómez 2020). Others, such as Calí mayor Jorge Iván Ospina, were trying to prevent them from even entering their cities and towns by proposing a "sanitary blockade" (Sánchez 2020) or by closing city borders. Still others offered to pay for their tickets back to Venezuela or refused to provide any resources for this population (Welsh 2020b; *El Tiempo* 2020b). Adding fuel to the fire, Bogotá mayor Claudia López Hernández reinforced stereotypes that Venezuelans are responsible for crime in the city by saying: "I do not want to stigmatize the Venezuelans, but there are some immigrants involved in criminality who are making our lives difficult. We welcome whoever comes to earn a decent living, but whoever comes to commit a crime should be deported without contemplation" (Pertuz 2020; Guzmán and Ponce 2020). All of this has fueled anti-Venezuelan xenophobia and mistaken ideas that they threaten Colombian society.

For Venezuelan women, rampant xenophobia combines with sexism, as they constantly endure sexual harassment and verbal abuse and are attacked as "dirty foreigners" (Zulver 2019). On top of that, they endure constant accusations of having too many children and coming to Colombia to have babies and access its welfare system. Accusations such as these came from journalist Claudia Palacios, who demanded that Venezuelan women stop having babies (Palacios 2019).

Similarly, the LGBT immigrant community has faced homophobia, xenophobia, and other prejudices based on gender identity. We saw one such example in Maicao Mayor Mohammed Dasuki's comments about gay prostitutes. More broadly, according to Wilson Casteñada, the head of Colombian LBGTQ rights NGO Caribe Afirmativo, "Xenophobia is combining with prejudices against gender identity. And in some cases, this elevated level of criminality is led by state authorities and can include practices like deportation" (Granados 2019).

In summary, confronting various forms of xenophobia has been part of the everyday experience of displaced Venezuelans in Colombia. They have been constantly accused of stealing jobs; being violent, dirty, and diseased; and threatening the stability of Colombian society. Worse still, these discourses and stereotypes led to the final aspect of the Venezuelan experience I will discuss: systemic violence.

Systemic Violence

> We have seen so many women go and never come back. Actually, a friend of mine went to the house of a man she did not know for a job [sex work]. He beat her and raped her and told her it was because she was Venezuelan and Venezuelans are worth nothing in Colombia.
>
> Diana (Personal correspondence)

> They broke our windows and stole our furniture . . . We might have to return to Venezuela by foot, and we might go hungry once we get there . . . But at least we will not be humiliated anymore.
>
> Jason Zabala, a Venezuelan coffee vendor (Rueda 2018)

Systemic violence occurs when "members of some groups live with the knowledge that they must fear random, unprovoked attacks on their persons or property, which have no motive but to damage, humiliate, or destroy the person" only because someone is a member of a certain group or nationality (Young 1990; Wolf 2020). These acts are not about the violence itself, but rather, they are aimed at making someone feel vulnerable and afraid because of their nationality. Again, this describes many experiences of Venezuelan immigrants in Colombia.

Anti-Venezuelan violence is ubiquitous in the border region. In part, this is because the Colombian government does not actually control large swaths of its border with Venezuela. Consequently, antigovernment and organized crime groups have filled the void and, as Human Rights Watch has documented, "have committed a range of abuses against civilians, including killings, disappearances, sexual violence, child recruitment, and forced displacement. There are also reports that they are planting antipersonnel mines" (Human Rights Watch 2019). Moreover, both "Venezuelan men and women are increasingly recruited—sometimes forcibly—into armed groups that operate on both sides of the border" (Zulver 2019). The women, in particular, are at risk of being kidnapped and are "required to cook, clean, harvest coca, and sexually service male combatants" (Zulver 2019). These groups monitor the *trochas* (or human-made informal crossings) where migrants cross illegally and charge men bribes. But "if a female migrant doesn't have money . . . sex is demanded as payment" (Zulver 2019). Similarly, "lesbian and bisexual women recounted attempts by drivers transporting them to abuse them sexually or even sell them to illegal armed groups" (Granados 2019).

Once the migrants get settled, violence remains a constant part of their daily lives. They face police brutality, physical violence, and threats of extortion and kidnapping from criminal groups. In late October 2018, for example, a mob attacked Venezuelan homes in Bogotá and beat one Venezuelan man to death in response to completely baseless rumors on social media that Venezuelan immigrants were kidnapping children (Rueda 2018). In the first week of the month of May 2019 alone, police confirmed that there were "three reported cases of transwomen shot by police in Valledupar, with separate reports in Bogotá of grave physical assaults on Venezuelan transwomen" (Granados 2019). More generally, on average, "one Venezuelan died violently in Colombia each day in the first eight months of 2019" (Reuters 2019), and "Colombia's Department of Legal Medicine stated Sunday that, between January and June 2019, 233 Venezuelans were killed in Colombian territory, including 206 men and 27 women" (Koerner and Vaz 2019), the majority of which were caused by gunshots or stabbings.

For Venezuelan women, the challenges are especially great (Maldonado, Londoño, and Ospina 2019). Venezuelan women endure harassment and sexual exploitation in exchange for food, water, medicine, personal hygiene products, and other basic goods. They also face such incidents at work or "when searching for work, [when] 'men trick women with job opportunities but in reality, they have intentions of some kind of exploitation'; 'other times women are drugged and taken advantage of'" (International Refugee Committee 2018). When this occurs, girls and women are often victimized by trafficking networks and/or are vulnerable to being recruited (or kidnapped) into sexual exploitation rings that target them (Maldonado, Londoño, and Ospina 2019). They are also very vulnerable to sexual assault and rape.

Unfortunately, there are numerous examples of sexual assault, rape, and torture against displaced Venezuelan women, but I will highlight just one. In September 2017, police officer Darwin Flórez Miranda was arrested for raping a 29-year-old Venezuelan woman, Iris, on her way home from work. As he raped her, he threatened her with deportation if she reported the crime (*Pulzo* 2017). Colombian journalist Claudia Ayola perfectly captured how this is global oppression in the form of systemic violence, as opposed to a deranged, random act committed by a "bad apple," saying: "This was not a common rape, its occurrence is deeply related to her status as an immigrant. The aggressor was a member of the police force of the receiving nation, an employee of the State, a police officer, an actor of the armed state executing abuse with the threat of deportation" (Ayola 2019).

Even more grave, the violence does not stop there. As Venezuelan sociologist Esther Pineda notes: "Venezuelan women in Colombia are increasingly victims of femicide" (Pineda 2019). Pineda reports that "between April 2018 and April 2019, at least 22 femicides were committed against Venezuelan women in Colombian territory" (Pineda 2019). Red Feminista Antimilitarista found that since early 2018, there have been at least two femicides of Venezuelan migrant women in Colombia per month, with no sign that the trend is abating (Observatorio Feminicidios Colombia 2019). Most recently, Francisco de Vitoria identified 83 cases of femicide against Venezuelan migrants in Colombia (Goldberg 2020). That is not only systemic violence but also violence against Venezuelan women *because they are Venezuelan migrant women* in order to make this group feel vulnerable and afraid at all times, as well as to send a message to other displaced Venezuelans that they are not welcome in Colombia.

Conclusion

As I have shown above, despite many of Colombia's efforts to welcome their Venezuelan brethren, the experiences of displaced Venezuelans in Colombia are filled with strife and danger. In part, this is due to the fact that, as we saw earlier, Colombia lacks resources to adequately meet their needs. It also owes to the fact that Colombia has not updated its legal framework to adequately respond to this population's needs or that it has continued to perceive Venezuelans as *temporary* residents of Colombia who will return once the problems in their country are resolved (thus leading Colombia's government and citizens to see their job as creating temporary programs rather than more permanent responses). In part, this could be because Colombia has received more credit than it deserves since it is compared to other nations in the Americas, like Peru, Chile, and Ecuador, that have been expressly hostile to this group (for example, imposing new strict entrance requirements that most Venezuelans cannot meet, effectively closing their borders to thousands of them) (Moloney 2019; Collins 2019b; Baddour 2018). Perhaps, in this way, Colombia's efforts have been overestimated.

Regardless of the reasons, the fact remains that global oppression is a constant in the lives of displaced Venezuelans in Colombia. Despite all the above, there is reason for hope. As more Venezuelans have been forced to flee, Colombia's president has continued to affirm the nation's commitment to help. Colombia introduced various plans for action in 2021, such as opening the application and renewal process for *Permiso Especial de Permanencia* through February 2021, when it was halted (El Venezolano Colombia 2020), introducing a six-point plan for helping Venezuelans during the COVID-19 pandemic (UNHCR 2020), and working with the United Nations to implement a regional plan for helping displaced Venezuelans in the region (Noticias ONU 2020). The city of Barranquilla won a grant to help Venezuelans find work in their city (Palomares 2021b), and Colombia's border manager has asked for Venezuelan degrees to be recognized so that more can find work in their fields (Palomares 2021a). Beyond this, scholars have begun to investigate the issue, and numerous nonprofit groups have been working to improve these circumstances, including but not limited to Project Hope, the International Refugee Committee, the South American Initiative, and the United Nations High Commissioner for Refugees. As one man told me on the bus, "I am here to find work and give my children opportunities . . . and I think, with time, I will." I hope he is right and that Dania Ortiz, Grebelin Muñoz, Gabriel Lugo, Mayra, Andrea, and so many more of their Venezuelan *hermanos* can achieve the life they have risked so much to attain.

Notes

1 Roughly 5 million (or 10 percent of Colombia's population) still reside abroad (LaRosa and Mejía 2017, 215).

2 For more detailed data, see https://data.unhcr.org/en/situations/syria.

3 While I do think this framework could be useful for explaining why and how other migrant groups face injustice as a result of being immigrants from specific nations, here my analysis is the experience of displaced Venezuelans.

References

Ajiaco, Ricardo. 2020. "Migrantes Venozolanos llegarían a 2 milliones en 2020." *El Tiempo*, January 22, 2020. https://www.eltiempo.com/politica/partidos-politicos/lo-restos-de-colombia-frente-a-una-migracion-venezolana-que-no-cesa-453616.

Arnson, Cynthia J. 2019. "The Venezuelan Refugee Crisis Is Not Just a Regional Problem: Latin American Neighbors Are Pulling More than Their Weight." *Foreign Affairs*, July 26, 2019. https://www.foreignaffairs.com/articles/venezuela/2019–07–26/venezuelan-refugee-crisis-not-just-regional-problem.

Ayola, Claudia. 2019. "Mujeres Migrantes." *El Heraldo*, October 27, 2019. https://www.elheraldo.co/columnas-de-opinion/claudia-ayola/mujeres-migrantes-675683.

Baddour, Dylan. 2018. "Ecuador Shuts Its Border to Venezuelan Refugees Amid Historic Exodus." *The Washington Post*, August 20, 2018. https://www.washingtonpost.com/world/

the_americas/ecuador-shuts-its-border-to-venezuelan-refugees-amid-historic-exodus/2018/08/20/28223fec-a48c-11e8-ad6f-080770dcddc2_story.html.
———. 2019a. "Colombia's Radical Plan to Welcome Millions of Venezuelan Migrants." *The Atlantic*. January 30, 2019. https://www.theatlantic.com/international/archive/2019/01/colombia-welcomes-millions-venezuelans-maduro-guaido/581647/.
———. 2019b. "This Country Is Setting the Bar for Handling Migrants: The Conservative Government of President Iván Duque in Bogotá Is Offering Citizenship to Colombian-Born Babies of Venezuelan Mothers." *The Atlantic*. August 16, 2019. https://www.theatlantic.com/international/archive/2019/08/colombias-counterintuitive-migration-policy/596233/.
Bartsfield, Jenny. 2019. "Colombia Gives Venezuela Newborns a Start in Life." United Nations Refugee Agency, October 14, 2019. https://www.unhcr.org/en-us/news/stories/2019/10/5da42be64/colombia-gives-venezuela-newborns-start-life.html.
Blu Radio. 2018. "Mientras el mundo cierra fronteras, Colombia opta por la fraternidad: Iván Duque." BluRadio. September 2. 2018. https://www.bluradio.com/nacion/mientras-el-mundo-cierra-fronteras-colombia-opta-por-la-fraternidad-ivan-duque.
Castillo Arenas, Gustavo A., and Patrick Ammerman. 2020. "A Country that Welcomes Migration." *Yes!* February 19, 2020. https://www.yesmagazine.org/issue/world-we-want/2020/02/19/colombia-venezuela-migration.
The City Paper. 2019. "Colombia Deports 59 Venezuelans for Acts of Vandalism during Bogotá Protests." November 25, 2019. https://thecitypaperbogota.com/news/colombia-deports-59-venezuelans-for-acts-of-vandalism-during-bogota-protests/.
CNN Español. 2019. "Denunciona amenazas en panfletos contra inmigrantes venezolanos en Colombia." August 1, 2019. https://cnnespanol.cnn.com/2019/08/01/denuncian-amenazas-en-panfletos-contra-inmigrantes-venezolanos-en-colombia/.
Collins, Joshua. 2019a. "Why You Don't Want to Be a Venezuelan Woman Right Now." *The New Humanitarian*. September 17, 2019. https://www.thenewhumanitarian.org/news-feature/2019/09/17/Venezuela-femicide-maternal-mortality-rates-rising.
———. 2019b. "Venezuelans Stranded as Ecuador Imposes New Visa Rules." Al Jazeera News, August 26, 2019. https://www.aljazeera.com/news/2019/8/26/venezuelans-stranded-as-ecuador-imposes-new-visa-rules#:~:text=Ecuador%20on%20Monday%20joined%20Peru,home%20to%20retrieve%20family%20members.
Cortés, Javier. 2019. "Bogota, pionera en atencion a población venezolana." Alcaldia de Bogota, December 16, 2019. https://bogota.gov.co/servicios/bogota-atiende-la-poblacion-venezolana.
El Espectador. 2019. "Sin patria, sin casa, sin escuela: el calvario de niños venezolanos en Colombia." October 10, 2019. https://www.elespectador.com/mundo/mas-paises/sin-patria-sin-casa-sin-escuela-el-calvario-de-ninos-venezolanos-en-colombia-article-885285/.
Frye, Marilyn. 1983. *The Politics of Reality*. California: The Crossing Press.
Goldberg, Beverly. 2020. "La mayoria de muertes de venezolanas en el país, entre 2018 y 2019, fueron feminicidios." *El Espectador*, January 29, 2020. https://www.elespectador.com/colombia/mas-regiones/la-mayoria-de-muertes-de-venezolanas-en-el-pais-entre-2018-y-2019-fueron-feminicidios-article-902062/.

Gómez, Camila. 2020. "Soachunos rechazan la construcción de albergue para venezolanos y colombianos en condición de calle, en el territorio municipal." Soacha Iniciativa Ciudadana, *La Voz de la Comunidad.* April 3, 2020. https://soachainiciativaciudadana.com/soachunos-rechazan-la-construccion-de-albergue-para-venezolanos-e-indigentes-colombianos-en-el-territorio-municipal/.

González, Ximena. 2019. "Expulsaron a otros seis venezolanos que buscarían afectar orden público en el paro." *La Republica*, November 16, 2019. https://www.larepublica.co/economia/expulsaron-a-otros-seis-venezolanos-que-buscarian-afectar-orden-publico-en-el-paro-2933731.

Granados, Marcela Osorio. 2019. "Double Risk for LGBT+ Venezuelan Migrants Crossing into Colombia." *Worldcrunch*, July 3, 2019. https://worldcrunch.com/culture-society/double-risk-for-lgbt-venezuelan-migrants-crossing-into-colombia.

Grattan, Steven. 2020. "Venezuelan Migrants Face Rising Xenophobia in Latin America." *The New Humanitarian.* February 13, 2020. https://www.thenewhumanitarian.org/news-feature/2020/02/13/Venezuelan-migrants-xenophobia-Latin-America.

Guzmán, Sergio, and Juan Camilo Ponce. 2020. "Hate against Venezuelans in Colombia Is a Ticking Time Bomb." *The Global Americans*, November 10, 2020. https://theglobalamericans.org/2020/11/hate-against-venezuelans-in-colombia-is-a-ticking-time-bomb/.

Human Rights Watch. 2019. "The War in Catatumbo." August 8, 2019. https://www.hrw.org/report/2019/08/08/war-catatumbo/abuses-armed-groups-against-civilians-including-venezuelan-exiles.

IACHR (Inter-American Commission on Human Rights). 2019. *Annual Report 2019: Chapter IV.B Venezuela.* http://www.oas.org/en/iachr/docs/annual/2019/docs/IA2019cap4BVE-en.pdf.

Infobae. 2020. "El 2021 traerá una nueva llegada masiva de migrantes venezolanos a Colombia, advierten ONGs." December 20, 2020. https://www.infobae.com/america/colombia/2020/12/20/el-2021-traera-una-nueva-llegada-masiva-de-migrantes-venezolanos-a-colombia-advierten-ongs/.

Integración Social. 2019. "Bogotá, pionera y referente en atención a migrantes venezolanos." December 26, 2019. https://www.integracionsocial.gov.co/index.php/noticias/116-otros/3641-bogota-pionera-y-referente-en-atencion-a-migrantes-venezolanos.

International Refugee Committee. 2018. "Needs Assessment Report: Venezuelan Migrants in Colombia." November 6, 2018. https://www.rescue.org/sites/default/files/document/3302/ircassessment-venezuelansincolombianov2018.pdf.

IOM. 2020. "Venezuelan Refugee and Migrant Crisis." https://www.iom.int/venezuelan-refugee-and-migrant-crisis.

Janetsky, Megan. 2019. "Here's Why Colombia Opened Its Arms to Venezuelan Migrants – Until Now." *Foreign Policy*, January 14, 2019. https://foreignpolicy.com/2019/01/14/heres-why-colombia-opened-its-arms-to-venezuelan-migrants-until-now/.

Koerner, Lucas, and Ricardo Vaz. 2019. "One Venezuelan Migrant Killed Every Day in Colombia." Venezuelanalysis, August 29, 2019. https://venezuelanalysis.com/news/14641.

Kurmanaev, Anatoly, and Jenny Carolina González. 2019. "Colombia Offers Citizenship to 24,000 Children of Venezuelan Refugees." *New York Times*, August 5, 2019. https://www.nytimes.com/2019/08/05/world/americas/colombia-citizenship-venezuelans.html.

LaRosa, Michael J., and Germán R. Mejía. 2017. *Colombia: A Concise Contemporary History*. 2nd ed. London: Rowman & Littlefield.

Louidor, Woodly Edson, Óscar Javier Calderón Barragán, Alejandra Castellanos Bretón, Silvia Carolina Leal Guerrero, and Paola Julieth Sierra Abril. 2019. *Por una Frontera Garante de los Derechos Humanos: Colombianos, venezolanos y niños en riesgo de apatridia en el Norte de Santander (2015–2018)*. Bogotá: Pontificia Universidad Javeriana.

Maduradas. 2018. "¡DURO! 5 testimonios de inmigrantes venezolanos que demuestran que salir de Venezuela 'no es tan fácil.'" March 16, 2018. https://maduradas.com/duro-5-testimonios-de-inmigrantes-venezolanos-que-demuestran-que-salir-de-venezuela-no-es-tan-facil/.

Maldonado, Carolina Triviño, Juliana Martínez Londoño, and Fernancy Falla Ospina. 2019. *Libro IVE: Migrantes Venezolanas en Colombia: Barreras de Acceso a la interrupción voluntaria del embarazo*. Bogotá: La Mesa por la Vida y la Salud de las Mujeres.

Memes Ran-Damn. *Facebook*. September 20, 2019. https://www.facebook.com/memesrandamn/posts/1154990624694804/.

Moloney, Anastasia. 2019. "Is South America Closing Its 'Open Door' on Venezuelans?" Reuters, August 8, 2019. https://www.reuters.com/article/us-venezuela-migration-analysis-idUSKCN1UY27D.

Noticias ONU. 2020. "La ONU lanza el Plan Regional 2021 para proteger a millones de venezolanos en América Latina." December 10, 2020. https://news.un.org/es/story/2020/12/1485402.

Observatorio Feminicidios Colombia. 2019. "Vivas Nos Queremos: Femicidios de Migrantes Venezolanas en Colombia Enero a Abril 2019." *Boletin Nacional Especial*, July 14, 2019. https://observatoriofeminicidioscolombia.org/index.php/seguimiento/boletin-nacional/391-feminicidios-de-migrantes-venezolanas-en-colombia-enero-a-abril-de-2019.

Organization for American States. 2020. *Venezuelan Migration and Refugee Crisis*. https://www.oas.org/fpdb/press/OAS_Dic20-crisis-of-Venezuelan-migrants-and-refugees-situation-report.pdf.

Otis, John. 2020. "Thousands of Migrants Head Back to Venezuela to Flee Colombia's COVID-19 Lockdown." National Public Radio, April 28, 2020. https://www.npr.org/2020/04/28/846945447/thousands-of-migrants-head-back-to-venezuela-to-flee-colombias-covid-19-lockdown#:~:text=Press-,Thousands%20Of%20Migrants%20Head%20Back%20To%20Venezuela%20To%20Flee%20Colombia's,Venezuelans%20are%20fleeing%20on%20foot.

Palacios, Claudia. 2019. "Paren de parir." *El Tiempo*, June 12, 2019. https://www.eltiempo.com/opinion/columnistas/claudia-palacios/paren-de-parir-columna-de-claudia-isabel-palacios-giraldo-374742.

Palomares, Milagros. 2021a. "Gerente de Fronteras aboga por la homologación de títulos." Proyecto Migración Venezuela, January 2, 2021. https://migravenezuela.com/web/articulo/gerente-de-fronteras-aboga-por-la-homologacion-de-titulos-/2387.

———. 2021b. "Con $565 millones, Barranquilla impulsará empleo a migrantes." Proyecto Migración Venezuela, January 7, 2021. https://migravenezuela.com/web/articulo/barranquilla-gano-un-fondo-para-promover-el-empleo-migrante/2398.

Pertuz, José Luis. 2020. "Claudia López propone deportar 'sin contemplación' a venezolanos que cometan crímenes en Bogotá." BluRadio, October 29, 2020. https://www.bluradio.com/blu360/bogota/no-quiero-estigmatizar-a-venezolanos-pero-unos-nos-estan-haciendo-la-vida-cuadritos-claudia-lopez.

Pineda, Esther. 2019. "Explotadas y asesinadas: la vulnerabilidad de las mujeres venezolanas." 2019. *El Espectador*, December 20, 2019. https://www.elespectador.com/colombia-20/analistas/explotadas-y-asesinadas-la-vulnerabilidad-de-las-migrantes-venezolanas-article/.

Plataforma de Coordinación Interagencial para Refugiados y Migrantes de Venezuela. 2022. "Refugiados y Migrantes Venezolanos en la Región." https://www.r4v.info/sites/default/files/2022-08/2022.08.R4V_R%26M_Map_Esp%28Note%29.pdf.

Pozzebon, Stefano. 2020. "Covid-19 Forced Tens of Thousands of Venezuelans to Go Home. But Crossing the Border Is No Easy Task." CNN, July 14, 2020. https://edition.cnn.com/2020/07/13/americas/latam-venezuela-migrants-cucata-intl/index.html.

Proyecto Migración Venezuela. 2019. "Aparecen panfletos amenazantes contra venezolanos en Boyacá." November 8, 2019. https://migravenezuela.com/web/articulo/aparecen-panfletos-con-amenazas-a-los-venezolanos-en-ventaquemada-boyaca/1566.

Pulgarín, Jovan. 2017. "Emigrar a Colombia, cerquita pero no es tan fácil." *Clímax*, July 26, 2017. https://elestimulo.com/climax/emigrar-a-colombia-cerquita-pero-no-es-tan-facil/.

Pulzo. 2017. "Detiene a expolicía que habría violado a venezolana en Sabanalarga (Atlántico)." September 29, 2017. https://www.pulzo.com/nacion/captura-expolicia-habria-violado-venezolana-PP357490.

Ramírez, Santiago, and Sophia Gómez. 2020. "Explotación al migrante: empleos que destrozan vidas en Colombia." *Revista Semana*. https://especiales.semana.com/desprotegidos-la-vida-de-los-migrantes-en-colombia/explotacion-laboral-al-migrante-venezolano.html.

Relief Web. 2020. *Situation Report: Venezuelan Migration and Refugee Crisis*. https://reliefweb.int/report/colombia/situation-report-venezuelan-migration-and-refugee-crisis-december-2020.

Restrepo, Vanessa. 2018. "La explotación laboral que padecen los venezolanos en Medellín." *El Colombiano*, July 4, 2018. https://www.elcolombiano.com/antioquia/venezolanos-en-medellin-padecen-explotacion-laboral-ED8948446.

Reuters. 2019. "Violent Deaths of Venezuelans in Colombia on the Rise: Report." https://www.reuters.com/article/us-venezuela-migration-colombia-idUSKCN1VF0PE/.

Revista Semana. 2018. "'Colombia optó por la fraternidad': Iván Duque sobre el éxodo de los venezolanos." September 2, 2018. https://www.semana.com/nacion/articulo/colombia-opto-por-la-fraternidad-ivan-duque-sobre-el-exodo-de-los-venezolanos/581791/.

Rivero, Pablo. 2019. *Yes, But Not Here: Perceptions of Xenophobia and Discrimination towards Venezuelan Migrants in Colombia, Ecuador and Peru*. Oxfam International. October 2019. https://oxfamilibrary.openrepository.com/bitstream/

handle/10546/620890/bp_yes_but_not_here_en_xenophobia-migration-venezuela-251019-en.pdf.
Rojas, Jonathan O. Lopez, and Deisy K. Castro Carpentiero. 2020. "Discriminación, estereotipos y prejuicios sobre las mujeres migrantes venezolanas en la ciudad de Bogotá." Universidad de la Salle.
Rueda, Manuel. 2018. "Venezuelan Migrant Killed in Mob Attack in Colombia Capital." Associated Press, October 30, 2018. https://apnews.com/774e7fe070654c54a2b43e35dbcc0855.
Sanabria, María Teresa Palacios, and Beatriz Londoño Toro, eds. 2019. *Migración y Derechos Humanos: El Caso Colombiano, 2014–2018*. Bogotá: Universidad del Rosario.
Sánchez, Gustavo. 2020. "Cali bloquería ingreso de migrantes, por el sur de la cuidad, para evitar fuente de contagio de Covid 19." Alcaldía de Santiago de Cali, April 21, 2020. https://www.cali.gov.co/gobierno/publicaciones/153135/cali-bloquearia-ingreso-de-migrantes-por-el-sur-de-la-ciudad-para-evitar-fuente-de-contagio-de-covid-19/.
Stott, Michael, and Gideon Long. 2020. "Venezuela: Refugee Crisis Tests Colombia's Stability." *Financial Times*, February 18, 2020. https://www.ft.com/content/bfede7a4–4f44–11ea-95a0–43d18ec715f5.
El Tiempo. 2018. "Así se vivía cuando la ola migratoria era de Colombia hacia Venezuela." February 11, 2018. https://www.eltiempo.com/mundo/venezuela/anteriormente-la-ola-migratoria-era-de-colombianos-hacia-venezuela-181258.
———. 2019. "Cuánto ganan, dónde, y qué tipo de empleo tienen los venezolanos." September 19, 2019. https://www.eltiempo.com/economia/sectores/salario-de-venezolanos-en-colombia-donde-trabajan-y-en-que-empleos-segun-dane-413732.
———. 2020a. "3 de cada 4 venezolanos trabajan en Colombia sin un contrato laboral: Cifras del empleo que ocupan, según estudio de la U. Externado. Menguó afluencia de empresarios." February 26, 2020. https://www.eltiempo.com/economia/sectores/realidad-laboral-de-venezolanos-en-colombia-466664.
———. 2020b. "En Antioquia pagarán viaje de venezolanos que quieren volver a su país." April 14, 2020. https://www.eltiempo.com/colombia/medellin/en-el-oriente-antioqueno-pagaran-viaje-de-venezolanos-que-quieran-volver-a-su-pais-484358.
UNHCR. 2020. "Colombia's 6-point Plan for Venezuelan Migrants during COVID-19." Global Compact on Refugees Digital Platform. https://globalcompactrefugees.org/article/colombias-6-point-plan-venezuelan-migrants-during-covid-19.
El Venezolano Colombia. 2020. "Más de 132 mil Venezolanos, Portadores del PEP, podrán renovar su documento a partir de hoy." December 21, 2020. https://elvenezolanocolombia.com/2020/12/mas-de-132-mil-venezolanos-portadores-del-pep-podran-renovar-su-documento-a-partir-de-hoy/#:~:text=A%20partir%20de%20hoy%20y,Colombia%2C%20Juan%20Francisco%20Espinosa%20Palacios.
Villavicencio Día a Día. 2019. "'Se paga a millón de pesos por veneco muerto': cruda amenaza en Boyacá." November 7, 2019. https://www.villavicenciodiaadia.com/se-paga-a-millon-de-pesos-por-veneco-muerto-la-cruda-amenaza-hacia-los-venezolanos-en-colombia/#:~:text=Email-,%E2%80%9CSe%20Paga%20A%20Mill%C3%B3n%20De%20Pesos%20Por%20Veneco%20Muerto%E2%80%9D%3A,Hacia%20Los%20Venezolanos%20En%20

Colombia&text=En%20la%20v%C3%ADa%20del%20Alto,por%20veneco%20 muerto%20por%20ratas.
Welsh, Teresa. 2020a. "$1.44B Needed for 2021 Venezuela Response." *Devex*, December 12, 2020. https://www.devex.com/news/1–44b-needed-for-2021-venezuela-response-98770.
———. 2020b. "To Stop Covid-19 Spread, Colombia Halves Venezuela Response Services." *DevEx*, March 17, 2020. https://www.devex.com/news/to-stop-covid-19-spread-colombia-halves-venezuela-response-services-96780.
Wolf, Allison B. 2019. "Dying in Detention as an Example of Oppression." *Hispanic/Latino Issues in Philosophy Newsletter of the American Philosophical Association* 19(1): 2–8.
———. 2020. *Just Immigration in the Americas: A Feminist Account*. London and New York: Rowman & Littlefield International.
———. 2021. "Immigration Injustice in Colombia: Beyond the Question of Borders." Border Criminologies blog. https://www.law.ox.ac.uk/research-subject-groups/centre-criminology/centreborder-criminologies/blog/2021/04/immigration.
Young, Iris Marion. 1990. *Justice and the Politics of Difference*. Princeton, NJ: Princeton University Press.
Zulver, Julia. 2019. "At Venezuela's Border with Colombia, Women Suffer Extraordinary Levels of Violence." *Washington Post*, February 26, 2019. https://www.washingtonpost.com/politics/2019/02/26/venezuelas-border-with-colombia-women-suffer-extraordinary-levels-violence/.
Zuniga, Mariana. 2018. "'I'm Stuck Here': The Desperate Search for a Passport in Venezuela." *Independent*, September 21, 2018. https://www.independent.co.uk/news/world/americas/venezuela-passport-leave-economy-maduro-colombia-border-immigration-a8548806.html.

Conclusion

Gerda Heck and Eda Sevinin

This edited volume brings together various critical interrogations of mobility and the politics of migration in the Global South. As the title, *Making Routes: Mobility and the Politics of Migration in the Global South*, suggests, the main point of departure was to delve into how mobility and the politics of migration unfold in certain geographies that are designated by mainstream theories as "outside the West/North." Although recent academic advances (see Gonzalez and Ocampo 2021; Fiddian-Qasmiyeh 2020; Fiddian-Qasmiyeh and Daley 2018; Short, Hossain, and Khan 2020) bear witness to the increasing interest in South-to-South migration or mobility and migration in the Global South, we still believe that this is a much-needed discussion. Besides the enduring intensity and volume of human mobility in the Global South, recent developments show us how urgent it is to have an empirically grounded and conceptually rich critical discussion about mobility within and through the Global South and on the conceptualization of the Global South.

When we first began to compile this book in 2019, displacement had already long been at the top of the global political agenda. Since then, conditions that force millions of people out of their homes have only deteriorated. In mid-2022, the world was swirling in various crises, which led to further forced displacement of millions of people. Due to the Russian invasion of Ukraine, within only a few days, the world saw one of the most rapid displacements in history. In 2021, the abrupt end to the 20-year-long US occupation and subsequent Taliban political takeover in Afghanistan exacerbated one of the most protracted refugee situations in the world. Although at a slower pace for the time being, sudden- and slow-onset climate-related events force millions of people to displace, leave their living spaces, and move toward more habitable areas. By May 2022, the number of displaced has already reached a record high level of 100 million, surpassing a "dramatic milestone" (UNHCR 2022).

Along with these events, the COVID-19 pandemic, at the end of its second year, has amplified global inequalities. According to the 2022 Oxfam International report, "millions of people around the world are facing a cost-of-living crisis due to the continuing effects of the pandemic and the rapidly rising costs of essentials, including food and energy" (Oxfam International 2022). Unsurprisingly, global inequalities at their extreme further fuel the likelihood of displacement for millions of people. Besides the escalation of the conditions of displacement, nation-states' attempts to halt and, if possible, reverse (undesirable) human mobility have also grown, multiplied, and been put into action in unexpected ways. For instance, in Turkey, the much-praised welcoming policy for Syrian refugees seems to have been replaced by not-so-welcoming policies such as resettling more than 1 million Syrian refugees in Syrian territories under the control of the Turkish army (Hacaoglu 2022). Similarly, the Lebanese government threatens to "expel Syrians from the country if the international community does not repatriate them" (MacGregor 2022). In both cases, the current uninhabitable condition of a great portion of war-torn Syria no longer seems to be a major international law concern for Turkey or Lebanon. Fueled by these unanticipated developments, such as the emerging global cost-of-living crisis, animosity toward forcibly displaced populations is rising even in contexts that have been known to be welcoming for refugees (at least for Syrians) in the last decade.

However, at the same time, we also see that people nonetheless continue *making routes* for themselves, crossing international borders and undoing barriers erected before them. As the title of the volume implies, "making routes" first refers to physical mobility through which people on the move find ways to relocate themselves in other parts of the world or away from or back to their birthplaces. Migration and displacement, as well as the struggle of migrating people to undo various border regimes, point to an always and already politicized struggle, a struggle exposing but also challenging how and to whom the face of the planet will be opened and closed. "Making routes," therefore, is a recognition of the politicized and contentious nature not only of border regimes all over the world but also of the human practice of freedom of mobility.

The politicized nature of border regimes or human mobility is by no means limited to the conflicts at the physical borders of states. It is born out of the long duration of global inequalities and power asymmetries that have resulted from centuries of socioeconomic, political, and military relations. The recent UNHCR Global Report estimates that in 2021, "two-thirds of refugees and displaced populations came from only five countries: Syria, at 6.8 million, followed by Venezuela, Afghanistan, South Sudan and Myanmar" (Hassan and Westfall 2022). The same report states that 83 percent of the displaced population is hosted by low- and middle-income countries (UNHCR 2022). Thus, it is generally recognized that the contexts in which displacement takes place and displaced populations are hosted are predominantly in the "Global South," i.e., the countries listed by UNHCR as "low- and middle-income countries."

As powerful as such reports from the pens of international organizations or humanitarian agencies may be, they too readily assume the causes and consequences of these facts, as well as the geographical demarcations between South/North or East/West. Mainstream studies on migration and displacement, particularly those rooted in and drawing upon Western-origin concepts (for critiques, see Chimni 1998, 2009; Arat-Koç 2020; Connell 2007, 2014; Anzaldúa 1987), also join such demarcations, reinforcing and reifying the differences between regions, countries, and policies while overlooking how entangled global(ized) histories are.

The climate crisis and climate displacement are perhaps the most striking examples that show us the need to question such geopolitical demarcations. The climate crisis has powerfully revealed the artificial character of borders as the actual scale of concern is no longer nation-states but planet Earth. Hence, any action to be taken must be at the planetary level. In many respects, the climate crisis can be seen as a crisis of global in/justice. The Global North states are disproportionately more responsible for causing existing climate change, which includes the increasing frequency and intensity of droughts, floods, storms, and the rise of sea levels disproportionately impacting the Global South. According to a recent study, states of the Global North were responsible for 92 percent of the excess of global emissions in 2015 (Hickel 2020), whereas, in 2020, more than 30 million people were displaced globally due to climate-related risks, most of them in three regions: Latin America, Africa, and Southeast Asia (Refugees International 2021, 12).

At the same time, states of the Global North continue to close their borders to people displaced and dispossessed by the climate crisis. In France, a Front National representative even boldly declared in 2019, "borders are the environment's greatest ally; it is through them that we will save the planet" (Mazoue 2019; also see Turner and Bailey 2021). On the other side of the Atlantic, more and more migrants from Central America make their way toward the US with the aim of leaving behind not only violence, poverty, and food insecurity but also natural disasters. However, thousands fail each year, not only at the increasingly fortified US–Mexico border but also at the Mexican border with Guatemala (Miller, Buxton, and Akkerman 2021). In November 2020, the hurricanes Eta and Iota triggered torrential rains, flash floods, landslides, and crop damage in Honduras, El Salvador, Guatemala, and Nicaragua. More than 7 million people were estimated to have been affected by these two hurricanes (Miao 2021). The adverse impact of catastrophes like these two hurricanes, but also preexisting effects of climate change, for instance, the significant increase in negative rainfall deviation within the last ten years in Honduras, has contributed to people's decision to leave their homes heading toward North America (Bermeo and Leblang 2021). While more than 1.5 million people tried to cross the border to the US in 2021, Border Patrol agents conducted more than 1 million expulsions and deportations (American Immigration Council 2022).

That said, although for several years experts have been demanding that "climate refugees" be granted the same protection status as those recognized under the 1951 Geneva Convention, until today, there exists no legal definition for "climate refugees," and they are not recognized as refugees under the UN Refugee Convention. The 1951 Refugee Convention narrowly defines "refugees" as persons who cannot return to their country of origin because of a well-founded fear of persecution "for reasons of race, religion, nationality, membership of a particular social group or political opinion" (UNHCR 2021). Accordingly, the UNHCR chooses to call climate refugees "persons displaced in the context of disasters and climate change" (UNHCR 2021). Consequently, being left without any legal status recognized by individual countries or by the UN, most climate refugees fall between these definitions when it comes to international protection.

Another controversial development that will probably be discussed for a long time as the epitome of embedded inequalities of global politics is the deal between Rwanda and the UK government. On Tuesday evening of June 14, 2022, the inaugural flight of the so-called UK government scheme to send people seeking asylum in the UK to Rwanda was stopped after an intervention by the European Court of Human Rights (ECHR). Two months earlier, the two governments had signed a highly controversial deal, which would allow the UK to relocate asylum seekers who entered the country undocumented to Rwanda (Nair 2022). Similar proposals from European countries have been repeatedly attempted in recent years. Already in 2003, then-British prime minister Tony Blair aimed at establishing reception centers in so-called third countries. In 2005, German Interior Minister Otto Schily took on the idea and suggested "asylum centers in North Africa" (Carrera and Guild 2017). More recent attempts of this kind came from Austria and Denmark. However, the novelty of this deal is that the asylum cases of those relocated would be processed by Rwandan authorities, and if they are granted asylum, they would be allowed to stay in Rwanda. The UK, in turn, was supposed to pay Rwanda $157 million (€144 million) in order to cover the expenses for economic development and growth and would pick up operational costs of the program, along with accommodation and integration expenses. In the case where Rwanda accepts more migrants than expected, the UK has committed to increase its payments (Krippahl 2022). Although the first deportation was halted, at this moment, it is unclear how it will proceed in the future. With its presence, however, the program has already inspired Denmark to make a similar deal with the Rwandan government. Even if the UK–Rwanda deal does not survive, it appears we are witnessing the next phase of Europe's externalization policies on the African continent. In fact, it is an advancement of the idea that states can pay to absolve themselves of the responsibilities they assumed under the 1951 Geneva Convention.

These two examples (and many more that are beyond the scope of this conclusion) attest not only to the impossibility of drawing rigid boundaries to divide the world into two but also to how deep-seated power asymmetries rooted in colonial and imperialist histories are concealed by reified differences between the Global South and the Global North. The climate change displacement reveals the violent consequences of not only the climate crisis but also the uneven and colonial distribution of such consequences. Much like the other externalization policies, the Rwanda deal shows how—as opposed to the mainstream assumption that forced displacement happens *due to* conditions of the countries of the Global South—people who are undesirably mobile are forcibly and violently displaced in the opposite direction to their intended route. Both, in this sense, reveal the "postcolonial condition" and "the question of the colonial heritage and its permanence" (De Genova 2016, 79).

This brings us to the discussion on the Global South. Given that we recognize how entangled geographies and their historical developments are when it comes to discussing the politics of mobility, we do not seek to give definitive descriptions of the Global South that can be given clear cartographic representations. If not a spatially bounded geography (but perhaps an assemblage of differentially positioned and internally and externally hierarchized geographies), what is meant by the Global South? How do we understand, conceptualize, and analyze what is the Southern, if that even exists? And, perhaps most importantly, if and why do we need to understand, conceptualize, and act upon what is the Global South? These questions guided the writing and editorial processes of this book and each chapter, bringing forth, implicitly or explicitly, a particular focus.

In this edited volume, instead of assuming ready-made demarcations, we offer a more nuanced and relational approach grounded upon the histories of regions and countries in question. Throughout various thematic chapters, we challenge dominant geographical distinctions and boundaries between the South and the North and focus on mutual interdependence between these two vaguely defined and mutable geographies. Therefore, the title of the book, *Making Routes*, goes beyond physical mobility: it also becomes a conceptual tool for us to think of alternative routes and conceptual channels to undo the deep-rooted disciplinary boundaries.

According to Siba Grovogui (2011, 177), "Global South is an idea and a set of practices, attitudes, and relations. It is a disavowal of institutional and cultural practices associated with colonialism and imperialism." Similar to Grovogui's (2011) conceptualization—yet in an inverted manner—Nicholas De Genova (2016, 77) puts Europe under the spotlight and says, "Indeed, if Europe remains an abstraction, it is crucial all the same to recognize it as a real abstraction, produced and continuously sustained by sociopolitical relations." Drawing on these scholars (and many more), we take it that the Global South and the Global North as they exist today are epistemic abstractions and inventions (Ndlovu-Gatsheni and Tafira 2018) that are very much in relation to and constitutive of each other. As such, they are grounded on colonial knowledge production regimes as much as colonial expansionist campaigns that include but are not limited to enslavement, conquest, military invasions, and capitalist exploitation (Ndlovu-Gatsheni and Tafira 2020; see also De Genova 2016; Davies and Boehmer 2018; Fiddian-Qasmiyeh and Daley 2018). As eloquently put by Ndlovu-Gatsheni and Tafira (2018, 127), "imperial reason and scientific racism were actively deployed in the invention of the geographical imaginaries of the Global South and the Global North." In other words, the Global South (much like the Global North) is an assemblage of various geographies that have been put under the same category by virtue of their relationship and position in racialized histories and globalization (De Genova 2016). It stands out less as a geographical entity than an epistemic one.

Seen from this perspective, human mobility and the politics of migration stand out as one of the major fields through which to display how the invention of the Global South is still at work in contemporary politics. The depiction and representation of Global South countries as the refugee-producing contexts that require Western/Northern support and interventions carry traces of "North-centrism" (Arat-Koç 2020) in forced migration studies. These countries are still portrayed as those that lack the resources and capacity to govern themselves (and the displaced populations, for that matter). According to such representations, countries of the Global South end up reproducing and reinforcing conditions of displacement of people who would set off to eventually set foot in the North. Sedef Arat-Koç (2020, 372) aptly states, "the issues and solutions offered are connected to the politically and geopolitically defined interests, priorities, and concerns of Northern countries: their 'security,' their sovereignty, their resources (or lack thereof), their policies and institutions, their 'culture,' their labor market needs and their 'refugee crisis.'" The condition of possibility of this North-centrism lies not only in the historical unfolding of international relations. It is also reinforced by the knowledge production practices that at times concealed the roots of North-centric categories and suppressed Indigenous knowledge about mobility and at other times imposed rigid legal and sociopolitical categories on people who are on the move.

This is, of course, a very broad topic that far exceeds the scope of this conclusion. There is an increasing body of scholarship attempting to develop a "Southern perspective" or a "perspective from the Global South" in a way to challenge the colonial epistemologies that center around the Global North or the West. What we wanted to emphasize here is that we have, with this edited volume, joined these attempts to undo these disciplinary problems and invited students of migration and mobility to have a more nuanced discussion about the politics of migration anywhere in the world, particularly in the Global South. We brought together various conceptual as well as empirically grounded analyses that put the Global South at the center of study. Although we are well aware that featuring scholars from the South or shedding an empirical light on the happenings within the Global South do not suffice to develop a Southern perspective, it might be seen as a good beginning.

That said, the final section of this conclusion is an open invitation to everyone to think of and act upon alternative forms of mobility that challenge accepted notions of possibility. For this reason, we wanted to take a brief look at longer-term developments as well as recent attempts at *making routes* by, with, and for peoples of the Global South.

Freedom of Movements?

Perhaps the most striking topic running through debates in migration studies is the unequal conditions of free movement depending on the region in which one was born. While citizens of the Global North inherit privileged access to international mobility, this is denied to a majority of citizens of the Global South. Attempts in the academy to move the debate toward a global right to freedom of movement are often presented as too utopian and somehow unrealistic. However, at the regional level, there have already been steps toward a political implementation of this in the past. Prominently discussed in migration studies has been the right to move and reside freely within the European Union.

Meanwhile, for some decades now, in many subregions of the Global South, regional integration and free mobility have been seen as a means to overcome the often artificial (colonial) boundaries that separate communities and hinder development. Accordingly, there have been various arrangements in different regions seeking ways for free movement, however, to a large extent, regionally bound.

In Latin America and the Caribbean, arrangements for the free movement of persons and labor have been developed within the framework of the Caribbean Community and Common Market (CARICOM) and the Andean Community (CAN). The most progressive plan for the free movement of people in Latin America so far is the Southern Common Market (Mercosur). With the implementation of the Residence Agreement for Nationals of the Member States of Mercosur and Associated States (RAM) in 2009, migrant rights and mobility within the region were tremendously improved. Citizens of a Mercosur member state can currently acquire a two-year temporary residence permit in another member state, which can be converted into a permanent residence permit at the end of this period upon proof of economic income (Brumat 2020). Meanwhile, in Asia, though on a much more restrictive and limited level, the ASEAN Agreement on the Movement of National Persons, signed in 2012, aims to pave the way to liberalize and facilitate the mobility of skilled labor within the region (International Women's Rights Action Watch Asia Pacific 2016).

In Africa, the widely known regional initiative is the ECOWAS Protocol on the Free Movement of Persons, Goods, and Services, adopted in 1979 by the 15-member Economic Community of West African States (ECOWAS), which provides for the free movement of all ECOWAS citizens within the region. In a similar vein, agreements like the Southern African Development Community's (SADC) Protocol on Facilitation of the Movements of Persons signed in August 2005 or the East African Community (EAC) signed in November 2009 are aiming at regional free movement zones (Okunade 2021). In 2016, the African Union declared to strive for a "borderless" Africa with seamless migration within the continent. Two years later, over thirty African states signed the Protocol for Free Movement of Persons, intending to facilitate the free movement of capital, goods, services, and people, in order to create employment and improve the standards of living on the continent (Hirsch 2022). So far, its implementation has been piecemeal and still faces major obstacles, among them xenophobic tendencies against migrants in some countries or governments privileging their own nation-state interests. Furthermore, in the face of militant groups such as Boko Haram, many African states view freedom of movement as a potential security threat (Okunade 2021). However, in the long run, a path to general freedom of movement could be taken via regional free movement zones. Existing free movement zones could be expanded, and new ones could be created and eventually merged.

In the Introduction and in some contributions to this volume (see Introduction; Kandilige and Yeboah), we argued that externalization policies deployed predominantly yet not exclusively by Northern countries hinder other forms of freedom of movement and mobility for the rest of the Global South. Although this is true to a considerable extent, it is by no means absolute or irreversible. In fact, any attempt—academic, political, or social—to challenge these policies remains a crucial task. This task emphatically involves unearthing the colonial and imperial nature of attempts at halting the mobility of a great majority of the world's population. In response to the European and North American endeavors to stem and contain movements from the Global South, the postcolonial philosopher Achille Mbembe (2018) suggests that the African continent should open up to itself. Referring to the continent's colonial and postcolonial history, he states that "African and diasporic struggles for freedom and self-determination have always been intertwined with the aspiration to move unchained" (Mbembe 2018). The next stage of Africa's decolonization, he demands, should involve granting mobility to all its people and reshaping the terms of membership in a political and cultural ensemble that is not confined to the nation-state. We can also project Mbembe's suggestion to many other regions and continents. But we can also take up his invitation to produce new knowledge and push discussions toward new ends.

References

American Immigration Council. 2022. "Rising Border Encounters in 2021: An Overview and Analysis." Factsheet. https://www.americanimmigrationcouncil.org/sites/default/files/research/rising_border_encounters_in_2021.pdf.

Anzaldúa, Gloria. 1987. *Borderlands/La Frontera: The New Mestiza*. San Francisco: Aunt Lute Books.

Arat-Koç, Sedef. 2020. "Decolonizing Refugee Studies, Standing up for Indigenous Justice: Challenges and Possibilities for a Politics of Place." *Studies in Social Justice* 14 (2): 371–90.

Bermeo, Sarah, and David Leblang. 2021. "Climate, Violence, and Honduran Migration to the United States." *Future Development*. Brookings. https://www.brookings.edu/blog/future-development/2021/04/01/climate-violence-and-honduran-migration-to-the-united-states/.

Brumat, Leiza. 2020. "Freedom of Movement of Persons in South America – an Overview." *Bpb: Bundeszentrale für politische Bildung*. April 16, 2020. https://www.bpb.de/themen/migration-integration/laenderprofile/english-version-country-profiles/306425/freedom-of-movement-of-persons-in-south-america-an-overview/#node-content-title-1.

Carrera, Sergio, and Elspeth Guild. 2017. "Offshore Processing of Asylum Applications Out of Sight, Out of Mind?" Commentary. Thinking Ahead for Europe. *CEPS*. January 27, 2017. https://www.ceps.eu/ceps-publications/offshore-processing-asylum-applications-out-sight-out-mind/.

Chimni, B.S. 1998. "The Geopolitics of Refugee Studies: A View from the South." *Journal of Refugee Studies* 11 (4): 350–74.

———. 2009. "The Birth of a 'Discipline': From Refugee to Forced Migration Studies." *Journal of Refugee Studies* 22 (1):11–29.

Connell, Raewyn. 2007. *Southern Theory: The Global Dynamics of Knowledge in Social Science*. London: Routledge.

———. 2014. "Using Southern Theory: Decolonizing Social Thought in Theory, Research and Application." *Planning Theory* 13 (2): 210–23.

Davies, Dominic, and Elleke Boehmer. 2018. "Postcolonialism and South-South Relations." In *Routledge Handbook of South-South Relations*, edited by Elena Fiddian-Qasmiyeh and Patricia Daley, 48–58. New York: Routledge.

De Genova, Nicholas. 2016. "The European Question: Migration, Race, and Postcoloniality in Europe." *Social Text* 34 (3): 75–102.

Fiddian-Qasmiyeh, Elena. 2020. "Introduction: Recentering the South in Studies of Migration." *Migration and Society: Advances in Research* 3: 1–18.

Fiddian-Qasmiyeh, Elena, and Patricia Daley, eds. 2018. *Routledge Handbook of South-South Relations*. New York: Routledge.

Gonzalez, Celeste Cedillo, and Julieta Espin Ocampo, eds. 2021. *Human Displacement from a Global South Perspective: Migration Dynamics in Latin America, Africa and the Middle East*. Cham: Palgrave Macmillan.

Grovogui, Siba N'Zatioula. 2011. "A Revolution Nonetheless: The Global South in International Relations." *The Global South* 5 (1), Special Issue: The Global South and World Dis/Order, 175–90.

Hacaoglu, Selcan. 2022. "A Million Syrian Refugees in Turkey to Be Returned Home, Erdogan Vows." *Bloomberg*, May 3, 2022. https://www.bloomberg.com/news/articles/2022–05–03/turkey-plans-return-of-a-million-syrians-as-refugee-critics-grow.

Hassan, Jennifer, and Sammy Westfall. 2022. "Ukraine War Pushes Global Displaced to Record High, UN Says." *The Washington Post*, January 16, 2022. https://www.washingtonpost.com/world/2022/06/16/refugee-displaced-ukraine-syria-afghanistan/.

Hickel, Jason. 2020. "Quantifying National Responsibility for Climate Breakdown: An Equality-based Attribution Approach for Carbon Dioxide Emissions in Excess of the Planetary Boundary." *The Lancet Planetary Health* 4 (9): 399–404. https://www.sciencedirect.com/science/article/pii/S2542519620301960.

Hirsch, Alan. 2022. "How Africa Can Promote the Continental Free Movement of People." Africa at LSE. Analysis and Debate. January 14, 2022. https://blogs.lse.ac.uk/africaatlse/2022/01/14/how-african-union-can-promote-continental-free-movement-of-people-migration-protocol/.

International Women's Rights Action Watch Asia Pacific. 2016. "The Missing Women: Implications of the ASEAN Integration on Women Migrant Workers' Rights." Kuala Lumpur: IWRAW Asia Pacific. https://16dayscampaign.org/wp-content/uploads/2019/01/missing-women.pdf.

Krippahl, Cristina. 2022. "Rwanda Vows to Resettle UK Asylum-Seekers despite Criticism." *Deutsche Welle*, June 14, 2022. https://www.dw.com/en/rwanda-vows-to-resettle-uk-asylum-seekers-despite-criticism/a-62130956.

MacGregor, Marion. 2022. "Lebanon Threatens to Expel Syrian Refugees." InfoMigrants, June 21, 2022. http://www.infomigrants.net/en/post/41356/lebanon-threatens-to-expel-syrian-refugees.

Mazoue, Aude. 2019. "Le Pen's National Rally Goes Green in Bid for European Election Votes." France24. April 20, 2019. https://www.france24.com/en/20190420-le-pen-national-rally-front-environment-european-elections-france.

Mbembe, Achille. 2018. "The Idea of a Borderless World." *Chronic Chimurenga*. October 16, 2018. https://chimurengachronic.co.za/the-idea-of-a-borderless-world/.

Miao, Hannah. 2021. "Climate Change Is a Major Factor behind Increased Migration at U.S. Southern Border, Experts Say." CNBC. https://www.cnbc.com/2021/04/18/us-mexico-border-climate-change-factor-behind-increased-migration.html.

Miller, Todd, Nick Buxton, and Mark Akkerman. 2021. *Global Climate Wall: How the World's Wealthiest Nations Prioritise Borders over Climate Action*. Amsterdam: Transnational Institute. https://www.tni.org/en/publication/global-climate-wall.

Nair, Parvati. 2022. "How the UK's Plan to Send Asylum Seekers to Rwanda Is 21st-century Imperialism Writ Large." *The Conversation*, April 22, 2022. https://theconversation.com/how-the-uks-plan-to-send-asylum-seekers-to-rwanda-is-21st-century-imperialism-writ-large-181501.

Ndlovu-Gatsheni, Sabelo J., and Kenneth Tafira. 2018. "The Invention of the Global South and the Politics of South-South Solidarity." In *Routledge Handbook of South-South Relations*, edited by Elena Fiddian-Qasmiyeh and Patricia Daley, 127–40. New York: Routledge.

Okunade, Samuel. 2021. "Africa Moves towards Intracontinental Free Movement for Its Booming Population." *Migration Policy*. January 21, 2021. https://www.migrationpolicy.org/article/africa-intracontinental-free-movement.

Oxfam International. 2022. "Profiting from Pain: The Urgency of Taxing the Rich amid a Surge in Billionaire Wealth and a Global Cost-of-living Crisis." OCHA reliefweb. May 23, 2022. https://reliefweb.int/report/world/profiting-pain-urgency-taxing-rich-amid-surge-billionaire-wealth-and-global-cost-living-crisis-enar.

Refugees International. 2021. *Task Force Report to the President on the Climate Crisis and Global Migration: A Pathway to Protection for People on the Move.* Washington, DC: Refugees International. https://d3jwam0i5codb7.cloudfront.net/wp-content/uploads/2023/03/FINAL-TaskForceReporttothePresidentontheClimateCrisisandGlobalMigration-July142C2021.pdf.

Short, Patricia, Moazzem Hossain, and M. Adil Khan, eds. 2020. *South-South Migration: Emerging Patterns, Opportunities and Risks.* New York: Routledge.

Turner, Joe, and Dan Bailey. 2021. "'Ecobordering': Casting Immigration Control as Environmental Protection." *Environmental Politics* 31 (1): 110–31.

UNHCR. 2021. "Climate Change and Disaster Displacement." https://www.unhcr.org/en-us/climate-change-and-disasters.html.

———. 2022. "UNHCR: Global Displacement Hits Another Record, Capping Decade-long Rising Trend." June 16, 2022. https://www.unhcr.org/news/press/2022/6/62a9d2b04/unhcr-global-displacement-hits-record-capping-decade-long-rising-trend.html.

Notes on Contributors

C R Abrar PhD is the Executive Director of the Refugee and Migratory Movements Research Unit. He taught International Relations at the University of Dhaka for four decades. His research interests include refugee movement, labor migration, and statelessness. Abrar played a major role in organizing young members of an Urdu-speaking, camp-dwelling community to reassert Bangladeshi citizenship in 2008. He moved the higher judiciary and secured the release of unlawfully detained Rohingya refugees and returned Bangladeshi labor migrants. He is a contributor to *The Daily Star of Dhaka*. Currently, he is chair of the Bangladesh Civil Society for Migrants.

David Bolaños Acuña is a Costa Rican journalist and cofounder of Interferencia, an investigative journalism project, and Doble Check, a fact-checking initiative, both at the University of Costa Rica (UCR). His work focuses on the social impact of public policy in Central America, poverty, human rights, and popular culture. The focus of his fact-checking activism is disinformation about public health and migration.

Danyel Ferrari is a PhD candidate at Rutgers University's School of Communication and Information. She earned her master's degree in gender studies at Central European University in 2016 with a focus on migration and has since presented at numerous media and migration conferences, including the Migration Research Center at Koç University (MiReKoc), Sabancı University, and the International Communication Association

(ICA). Her dissertation focuses on empathy discourses on social media and public facing transnational artworks that address migration and displacement. In her research, she contextualizes these artworks through the political economies that produce and disseminate them, analyzes their use of affects and their public performance both on-site and on social media, and questions to what political ends and to whose benefit those affects are working.

Elena Habersky is a project manager at the Center for Migration and Refugee Studies, the American University in Cairo. She holds an MA in migration and refugee studies from the AUC and a BS in international studies from the University of Scranton. She has spent the better part of the last decade working and conducting research with African migrants and refugees in Amman, Jordan; Cairo, Egypt; and Kampala, Uganda. She has previously worked with JRS Jordan, Collateral Repair Project, and IOM Egypt and has been published in *America Magazine*, *The Cairo Review of Global Affairs*, *Egyptian Streets*, *Muftah Magazine*, and the *Oxford Monitor of Forced Migration*, to name a few. In 2020, she was the corecipient of the Moira O'Donnell Emerging Leader for Justice Award from the Ignatian Solidarity Network for her work with migrants and refugees.

Gerda Heck PhD is an assistant professor in the Department of Sociology, Egyptology and Anthropology and the Center for Migration and Refugee Studies, the American University in Cairo. Her academic work and research focus on migration and border regimes, urban studies, transnational migration, migrant networks and self-organizing, religion, and new concepts of citizenship. She has conducted research in Ghana, Germany, Brazil, China, Democratic Republic of the Congo, Egypt, France, Morocco, Turkey, and the USA. From 2010 to 2013, she was a postdoctoral fellow in the international and interdisciplinary research project Global Prayers – Redemption and Liberation in the City. In 2016, she conducted research in Turkey within the scope of the international research project Transit Migration 2: A Research Project on the De- and Re-Stabilizations of the European Border Regime. From September 2021 to January 2022, she was a visiting scholar within the scope of the BECHS-Africa Fellowship at the Institute of African Studies at the University of Ghana.

Evrim Hikmet Öğüt completed her PhD in ethnomusicology at Istanbul Technical University Centre for Advanced Studies in Music with a dissertation titled "Music in Transit: Musical Practices of the Chaldean-Iraqi Migrants in Istanbul." She teaches as an associate professor at Mimar Sinan

Fine Arts University, Ethnomusicology Program. Her ongoing research in Istanbul focuses on the musical practices of Syrian musicians. Her related video interview series is available on www.soundsbeyondtheborder.org. Besides her studies in Istanbul, since 2020, she has been researching the Arab identity's re-formation through music in New York as a visiting scholar at the City University of New York.

Leander Kandilige PhD is a Senior Lecturer in Migration Studies at the Centre for Migration Studies, University of Ghana. His research interests include migration policy development; theories of migration; migration, poverty and inequality; labor migration; and migration and development. He is a Country Lead for Ghana on the MIGNEX Project and a researcher in the MIDEQ Project. His research has been funded by several international organizations, including the EU, German Development Institute (DIE), GIZ, DFID, ICMPD, IOM, and ILO/OECD.

Mélanie Montinard graduated with a master's degree in international law from the universities of Strasbourg (France) and Münster (Germany). She is a professor of social anthropology at the National Museum, Federal University of Rio de Janeiro (UFRJ-Brazil), whose lines of research focus on the dynamics of Haitian networks and mobility in the Americas. Since 2017, she has served as the cofounder of Mawon, a Brazilian association that promotes mobility, diversity, connections, and integration of migrants, together with UN partners and the private sector. Currently, she is professor of intercultural communication and transnational migration at UFRJ.

Duduzile Ndlovu is a postdoctoral fellow at the African Centre for Migration and Society, University of the Witwatersrand. Her research focuses on migration in the Global South using arts-based research methods as a form of decolonizing knowledge production. She wrote her PhD dissertation on Zimbabwean migrants' memorials of Gukurahundi violence in Johannesburg, using performance, poetry, music, and film, and translated the thesis into poetry to access a wider and nonacademic audience. She was awarded the Newton Advanced Fellowship (2018–20) at the Centre of African Studies, University of Edinburgh.

Sara Sadek PhD is an expert on migration issues in the Middle East and North Africa regions with a doctorate in political science from the University of York, UK, and a master's degree in refugee studies from the University of

East London, UK. Sadek has served as an adjunct professor at the Center for Migration and Refugee Studies at the American University in Cairo. Since 2004, Sadek has been engaged in migration issues as a researcher, trainer, and consultant for key international organizations and academic institutions: International Organization for Migration (IOM), United Nations High Commissioner for Refugees (UNHCR), the United Nations Economic and Social Commission for West Asia (UN-ESCWA), the Spanish Cooperation in Egypt (SC), the Danish Refugee Council (DRC), and Center for Migration and Refugee Studies/American University in Cairo (CMRS/AUC), Duke University, Oxford University, and others.

Carlos Sandoval-García obtained his PhD in cultural studies from the University of Birmingham, UK. He is a professor in the Media Studies School at the University of Costa Rica. His monograph *Exclusion and Forced Migration in Central America: No More Walls* (Palgrave Pivot, 2017) was the starting point of the documentary *Home in a Foreign Land*, available online with subtitles in a number of languages. In 2020, Sandoval-García published *Centroamérica desgarrada: Demandas y expectativas de jóvenes residentes en colonias empobrecidas* (CLACSO). Currently, he coordinates a PhD program on Central America and is active in a number of networks of solidarity regarding migrants' rights in Costa Rica.

Eda Sevinin received her PhD in 2022 from the Department of International Relations at Central European University, Austria. Her PhD project focused on the Islamic humanitarian networks working with refugees in Turkey.

Tasneem Siddiqui is a professor of political science, University of Dhaka, and founder and chair of the Refugee and Migratory Movements Research Unit (RMMRU). She has extensively published on forced and voluntary migration and climate change-related migration issues. She led the drafting of the National Strategy for Internal Displacement in Bangladesh 2021, the Overseas Employment Policy 2006, and was a member of the committee that drafted the Overseas Employment and Migrants Act of 2013. She is on the Global Editorial Board of the *Oxford Journal of Migration Studies* and a member of the Advisory Committee of the Platform on Disaster Displacement (PDD) and the International Displacement Monitoring Centre.

Sally Souraya is a Lebanese visual storyteller and performer living in London. She has a particular interest in the artistic process and how it unfolds as a work in progress. At the intersection between art, anthropology, and activism, Sally's creative practice adopts an experimental approach rooted in observations, words, and movements. Her work aims to raise awareness and promote social change and well-being. She has performed and exhibited her work in different countries, including Lebanon, Belgium, the United Kingdom, China, and Italy.

Kudakwashe Vanyoro is a lecturer at the Department of Anthropology, University of the Witwatersrand, Johannesburg, and his research interests are migration, temporality, borders, humanitarianism, and governance in Africa. His doctoral research explored how temporal disruptions at international borders shape experiences and modes of waiting of irregular Zimbabwean migrants at the Zimbabwe–South Africa border who have arrived in South Africa but are restricted in moving further into the interior. Through this inquiry, his work reveals how waiting is a component of both governing Zimbabwean migrants and seeking agency through the relationship between time, space, and humanitarianism.

Allison B. Wolf PhD is an associate professor of philosophy and affiliated faculty member in the Center for Migration Studies at Universidad de los Andes in Bogotá, Colombia, where she teaches political philosophy, philosophy of immigration, and feminist philosophy. She is the author of *Just Immigration in the Americas: A Feminist Account* (Rowman and Littlefield International, 2020) and coeditor of *Incarnating Feelings, Constructing Communities: Experiencing Emotions in the Americas through Education, Violence, and Public Policy* with Ana María Forero Angel and Catalina González Quintero (Palgrave Macmillan, 2020). She is currently working on issues around immigration justice in a South-South context in the Americas.

Thomas Yeboah currently serves as research fellow of the Bureau of Integrated Rural Development (BIRD), Kwame Nkrumah University of Science and Technology, Ghana. He holds a PhD in Development Studies from the University of Cambridge, UK, and has nearly a decade-long experience working on the relationship between migration and development, youth livelihoods/employment, and sociocultural underpinnings of children's work. His research also focuses on youth migration, including migration journeys, decision-making, and lived realities in the realm of social and spatial mobilities, as well as how young people engage in the rural economy and its potential to provide decent and sustainable employment.

Printed in the USA
CPSIA information can be obtained
at www.ICGtesting.com
JSHW011037170224
57510JS00003B/5/J